Free-Space Optics: Enabling Optical Connectivity in Today's Networks

Heinz Willebrand, Ph.D., and Baksheesh S. Ghuman

201 West 103rd St., Indianapolis, Indiana 46290 USA

Free-Space Optics: Enabling Optical Connectivity in Today's Networks

Copyright © 2002 by Sams Publishing

International Standard Book Number: 0-672-32248-x

Library of Congress Catalog Card Number: 2001089233

Printed in the United States of America

First Printing: December 2001

04	03	02	01		4	3	2	1

Trademarks

Warning and Disclaimer

ASSOCIATE PUBLISHER
Linda Engelman

ACQUISITIONS EDITOR
Dayna Isley

DEVELOPMENT EDITOR
Laurie McGuire

MANAGING EDITOR
Charlotte Clapp

PROJECT EDITOR
Carol Bowers

COPY EDITOR
Karen A. Gill

INDEXER
Tim Tate

PROOFREADER
Jessica McCarty

TECHNICAL EDITORS
Peter Schoon
David Lively

TEAM COORDINATOR
Lynne Williams

INTERIOR DESIGNER
Gary Adair

COVER DESIGNER
Gary Adair

PAGE LAYOUT
Octal Publishing, Inc.

Contents at a Glance

Table of Contents

Preface

To the best of our knowledge, no one to date has written a comprehensive book about free-space optics (FSO). Until recently, it was considered to be a niche technology. However, due to the drastic changes in the communication network infrastructure, especially during the past couple of years, FSO might very well become a mainstream technology in the local loop access market. As carriers begin to adopt it, FSO is well on its way toward becoming a mainstream technology that not only addresses access applications, but also will play a major role in core networking applications. This is one reason we thought it was finally time to write a book about this exciting technology. We believe that FSO will be one of the most unique and powerful tools to address connectivity bottlenecks that have been created in high-speed networks during the past decade due to the tremendous success and continued acceptance of the Internet. Clearing these bottlenecks is crucial for the future growth and success of contemporary Internet society.

The Internet revolution goes hand in hand with the so-called "Computer Revolution" that started around the same time. The computer industry has balanced the development of more complex, powerful, and user-friendly application software with the development of more powerful hardware. In contrast, the telecommunications industry lives under different boundary conditions. Although the performance of computers followed Moore's law of increasing capacity, the existing local loop telecommunications infrastructure, or the "last mile" access infrastructure, simply was not able to keep up with the increasing bandwidth requirements of the Internet. Internet traffic actually scaled even faster than Moore's law. Because of the bandwidth limitations of the existing copper infrastructure, telecommunications providers (and ultimately their customers) were hitting the "telecommunications copper bandwidth wall." Even worse for the end user is the fact that Moore's law is still valid. This is in addition to the fact that the speed of in-building LAN networks is still drastically increasing, with no end in sight.

In anticipation of growing global bandwidth demand, telecommunication service providers have drastically increased their long-haul fiber network bandwidth capacities. Faster electronics in conjunction with wavelength division multiplexing (WDM) technology opened the real bandwidth potential of optical fiber, establishing it as the ultimate medium for multiterabit capacity long-haul backbone networks. The relaxed financial market of the past couple of years also provided an enormous amount of capital influx that was required to build up fiber connections between major metropolitan cities and high-capacity fiber access rings around these cities. However, even with the enormous amount of spending, reliable sources report today that only 5% of the commercial buildings in major metropolitan cities are connected to a fiber network. This is certainly not a high percentage figure for the amount of capital spent. In addition, this figure is far off the original expectations for fiber-based connectivity projected in the early 1990s. The 5% figure provides an indication of how long it takes and how expensive it is to lay fiber in the local loop.

In the aftermath of the recent market boom, capital spending to build out optical capacity in fiber networks has been largely reduced, although demand for inexpensive high bandwidth on the end user side is still high and will most likely increase continuously in the near future. Optical bandwidth is required to satisfy this demand. However, fiber deployment in the access and edge markets simply cannot be justified on a large scale due to the enormous capital upfront expenses. The next phase of the Internet service rollout is already on the horizon, driven by the ever-increasing capabilities of computer equipment, LAN performance, and software applications. This next phase entails the multimedia-driven society that will include high-bandwidth applications such as downloading movies and online bidirectional videoconferencing. Optical capacity in the access and edge networks will be needed to satisfy these demands.

We believe that FSO will become a crucial tool in the toolbox of service providers to bridge the gap between the end user and the high-capacity fiber infrastructure already in place. The inherent synergy between optical fiber and FSO will enable the transition from the old copper-based electrical society to the new optical society that uses light as transport media. We felt it was time to educate the market and users about the exciting potential of FSO to reach this goal. That's why we wrote this book. We hope that it helps you better understand the technology and its applications.

—Heinz Willebrand, Ph.D.
Baksheesh S. Ghuman

About the Author

Heinz Willebrand is the founder and chief technology officer of LightPointe. He studied physics at the University of Muenster and received his MS and Ph.D. from the Institute of Applied Physics in 1988 and 1992, respectively. In 1994, Heinz moved to the United States and held research positions at the University of Colorado in Boulder, Colorado, working on several DARPA- and NSF-sponsored projects related to fiber-optics communications and free-space optics. Heinz was also a technical advisor in the Global Telecommunications Management Program at the University of Colorado, where he taught classes in fiber optics and wireless communications technologies.

In 1995, Heinz co-founded Eagle Optoelectronics, Inc., as a small research company with the goal to commercialize fiber optics and free-space optics communication systems. The company attracted research and development funding from several government agencies. In 1998, Heinz spun off LightPointe from Eagle Optoelectronics with the goal to solely focus on the development and commercialization of high-capacity free-space optics systems.

Heinz is well known in the free-space optics and fiber-optics communities. He holds several patents and has been an invited speaker at many conferences covering various technical and business-related aspects of free-space optics. Heinz is considered one of the leading forces behind the recent success of free-space optics in the telecommunication industry.

Baksheesh S. Ghuman is a technology marketing expert and was most recently the chief marketing officer of LightPointe. At LightPointe Baksheesh covered three functional areas: product management, market communications, and market management. He was responsible for executing LightPointe's global marketing strategy, ensuring LightPointe's continued growth and leadership in free-space optics. A technologist by training, Baksheesh has more than 12 years of marketing experience in the field of telecommunications. He has held similar positions at Sorrento Networks, a manufacturer of optical networking equipment, and Electric Lightwave, a regional service provider. Baksheesh's experience has spanned a variety of roles in the fields of access, transport, switching, and management in metropolitan optical networking for carriers, CLECs, ISPs, RBOCs, and other service providers and vendors.

Baksheesh holds a Bachelor of Science degree from Arya College, Punjab University, and a diploma in systems management from the National Institute of Information Technology, New Dehli. Baksheesh has both an MS in Telecommunications and an MBA with a Marketing concentration from Golden Gate University, San Francisco. He has also attended executive management programs at the Stanford University graduate school of business. In addition, Baksheesh has published several articles in leading telecommunications magazines.

About the Contributor

Peter Schoon is the president and founder of System Support Solutions, Inc., the first specialty free-space optics integrator in North America. He holds a degree in financial management, certifications from Cisco, Citrix, Compaq, and Optical Access, and has done postgraduate work in corporate management and marketing. A driving passion for seeking out the new and better-faster-cheaper, while adhering to a no-nonsense approach to the fundamentals of good business practices, has attracted Peter to the emergent free-space optics industry.

Peter served as both a contributor and technical editor for this book.

About the Technical Editor

David Lively is a senior manager of market development for Cisco's Optical Networking Group. His teams are focused on developing strategies for the metro optical market to help service providers increase revenues and profitability. Prior to his current role, he worked in the DSL Business Unit where he was responsible for the integrated access and voice-enabled DSL product lines. David often speaks at conferences and seminars throughout the world on topics ranging across much of the service provider market, including the metro optical market, dial access, DSL, packetized voice, and enabling broadband applications for consumers. He holds a Bachelor of Science degree in Computer Engineering from Virginia Tech.

Dedication

*To my wife Milen, for her love, patience, and
constant support to fulfill a personal dream.*

—Heinz Willebrand

*I would like to dedicate this book to two men in my family: to my late grandfather,
Sardar Thakkar Singh Ghuman, whose sacrifices and hard work have inspired me; and also
to my father, Sardar Lakhbir Singh Ghuman, whose spirituality has taught me to be forgiving and
accepting and continues to guide me in all aspects of my life.*

—Baksheesh S. Ghuman

Acknowledgments

This book would not have been published without the help of the LightPointe team.

We would like to thank Jeff Bean and Kathleen E. Dana from the marketing team and Avtar Singh and Leigh Fatzinger from the business development team for their help with the business-related chapters.

We would especially like to mention Jerry Clark, Laurel Mayhew, Brian Neff, Cathal Osclai, and Bryan Willson from the engineering team for their support and the hours they spent after a regular working day to make this book happen.

We greatly appreciate the assistance of Hasan Imam of Thomas Weisel Partners, formerly of DLJ; Douglas Peterson of Pioneer Consulting; Helena Wolfe; Kerri J. Altom; and countless friends. Special thanks to Dayna Isley of Sams Publishing whose patient drive and spirit really guided us to complete the book on time. We would also like to thank Laurie McGuire, Carol Bowers, Karen Gill, and the rest of the Sams team for their hard work and support throughout this process.

Last but not least, Heinz Willebrand would like to thank Dr. Erhard Kube for the endless hours of discussions that made him understand many technical aspects of free-space optics. You are a great friend and teacher.

Tell Us What You Think!

As the reader of this book, *you* are our most important critic and commentator. We value your opinion and want to know what we're doing right, what we could do better, what areas you'd like to see us publish in, and any other words of wisdom you're willing to pass our way.

As an associate publisher for Sams, I welcome your comments. You can fax, e-mail, or write me directly to let me know what you did or didn't like about this book—as well as what we can do to make our books stronger.

Please note that I cannot help you with technical problems related to the topic of this book, and that due to the high volume of mail I receive, I might not be able to reply to every message.

When you write, please be sure to include this book's title and author as well as your name and phone or fax number. I will carefully review your comments and share them with the author and editors who worked on the book.

Fax: 317-581-4770

E-mail: feedback@samspublishing.com

Mail: Linda Engelman
       Sams Publishing
       201 West 103rd Street
       Indianapolis, IN 46290 USA

Introduction to Free-Space Optics

IN THIS CHAPTER

The demand for high bandwidth in metropolitan networks on short timelines is increasing. Further, requirements of flexibility and cost effectiveness of service provisioning (some connections could be temporary, whereas others are long term) have caused an imbalance. This problem is often referred to as the "last mile bottleneck." That terminology incorrectly limits the problem. Similar issues exist in various parts of metropolitan networks—in their core, access systems, and their edge. Instead of calling it the "last mile bottleneck," the appropriate name for it should be the "connectivity bottleneck." This title addresses the overall problem more accurately. This issue is an issue not just in the last mile only, but as many network planners can attest, it is everywhere in metropolitan networks.

A few alternatives are available to address this "connectivity bottleneck" from a technology standpoint. Whether they make economic sense is another issue entirely. This chapter briefly examines each alternative, and then focuses a bit more on the potential benefits of free-space optics as a solution to the connectivity bottleneck.

Alternative Bandwidth Technologies

The first most obvious choice for addressing the bandwidth shortage is fiber. Fiber, without a doubt, is the most reliable means of optical communications so far, but the digging, cost to lay that fiber, and time to market are the most prohibitive factors. In addition, after you lay fiber, it becomes a sunk cost; if the customer leaves, it becomes almost impossible to recover that cost. Even though fiber is technologically superior to free-space optics, it is significantly more costly.

A second choice is radio frequency (RF). This technology is mature and has been deployed. RF-based networks require immense investments to acquire the spectrum licenses, yet they cannot scale to optical capacities. (The ceiling today is 622 Mbps.) However, RF-based networks can go longer distances. When compared to free-space optics, RF does not make economic sense.

A third alternative is all the copper-based technologies, such as cable modems, T1s, and DSLs. Even though copper infrastructure is available on a wider scale, and the statistic of buildings connected to copper is much higher than fiber, it is not a viable alternative for solving the connectivity bottleneck. The problem is bandwidth scalability. Copper's distance per segment and throughput is inherently limited; therefore, its potential to solve the connectivity bottleneck is also limited. And just as serious a problem for copper is that it is owned by the Incumbent Local Exchange Carriers (ILECs). This in all likelihood means that the cost per Mbps will remain high, at least for the foreseeable future.

The fourth, most viable alternative is free-space optics. FSO represents the most optimal solution in terms of technology (optical), bandwidth scalability, speed of deployment (hours versus months) and cost effectiveness (at least one fifth).

It is a well-known statistic in the telecommunications industry that only 5% of the commercial buildings in the United States are connected to a fiber backbone, yet 75% are within 1 mile of fiber [1]. Each building within 1 mile has four additional buildings within 100 meters (m) of it. Presumably, these businesses run high-speed LANs and would find it quite frustrating to be connected to the outside world through low-speed connections such as DSL, cable modems, or T1s.

Most of the trenching to lay fiber has been done for improving the metropolitan core (backbone), whereas the access and edge have completely been ignored. This situation has given birth to "an optical dead zone"—the complete optical disconnect between the core and the edge + access. Studies have shown that such disconnects also occur within the core, primarily due to cost constraints combined with moratoriums, nonscalable technologies that are deployed (such as LMDS), time-to-market, and so on. Metropolitan optical networks have not yet delivered on their promise: High capacity at affordable prices still eludes the ultimate end user. This is where free-space optics makes its entry.

Free-space optics has offered service providers with a viable alternative or complement to fiber optics for optical connectivity. In comparison to RF, the incumbent "alternative access" technology, FSO brings lower cost, higher-bandwidth, security, flexibility, and reduced time-to-market.

The industry has a misconception or lack of awareness about free-space optics; consequently, FSO has been classified as a wireless technology when it is clearly an optical technology. The distinction is in FSO's high optical transmission abilities and its lack of need for spectrum licenses.

> **NOTE**
>
> The *core* of the network refers to the backbone network, the network that supports the high-bandwidth transport. It also connects major traffic hubs or central offices (COs) of telephone companies.
>
> *Access* implies that part of the network that enables a business to access the core of the network. It connects the central offices of the telephone company to customer premises.
>
> *Edge* is the network that is within a building, campus, or LAN.

Fiber Versus FSO

Because FSO and fiber optics enable similar bandwidth transmission abilities, it is important to compare them. One of the most important points of comparison and contrast between them is in the way they transmit light. Light can be transmitted either through free space or a confined medium.

Fiber-Optic Cable

The most common and well-known confined medium is fiber-optic cable. A fiber-optic cable carries a light signal from point A to point B, but the light signal must be generated first. Light sources are devices that generate the light in optical networks.

A light source converts an electrical signal carrying voice/data/video content into an optical signal. The process by which the electrical signal is mapped onto the optical signal is called *modulation*. The light source can perform the modulation (self-modulation) or do so with the aid of an external shutter, called a *modulator*. For a digital signal—that is, a stream of 0s and 1s—modulation can be achieved by simply turning the light source on and off in response to an electrical 1 or 0. Think of a torch light analogy. You can communicate with someone far away during a dark night by turning a torch light on and off in some predetermined sequence. In modern optical communications, the laser plays a role similar to the torch.

Light sources used in an optical network must possess certain characteristics. These may vary depending on a variety of factors, such as the type of fiber used, the data rate, and cost. In general, most optical communication applications require light sources that possess a number of key characteristics:

- Brightness: All other factors being equal, the brighter the light coming out of the source, the farther it can travel through the fiber before requiring amplification/regeneration, and the more cost efficient the transmission becomes. Given the high cost of light amplifiers and regeneration equipment, bright light sources are imperative in transmission systems. In technical terms, this means that the light source must have high flux or radiant power. It must emit many photons within a narrow band of wavelengths.

- Highly focused: The core of the fiber that carries the light is extremely small—less than the diameter of your hair. If the light beam from the source diverges too quickly, most of the light will not enter the fiber core and will be wasted. Thus, the area over which the light source emits must be small compared to the area of the fiber core.

- High modulation speeds: In directly modulated applications, where the light source turns on/off in response to the incoming electrical signal, the light source must be able to do so at speeds of millions/billions of times per second. In externally modulated applications, in which a high-speed shutter is placed in front of the light source, this property is not critical.

- Wavelength matching with fiber: As Chapter 3, "Factors Affecting FSO," discusses in more detail, in certain wavelengths, light suffers the least loss in a fiber medium; these are called *transmission windows* of the fiber. To maximize the distance that light can travel through fiber, the light source must emit at wavelengths within the transmission window of fiber.

- Reliable: In today's communications systems, a single strand of fiber carries millions of telephone calls and other mission-critical data. A failure of the light source can terminate all these calls and halt the transmission of mission-critical data. Reliability of the light source is critical. In undersea networks, where a repair trip can be expensive and time consuming, any deployed device must pass the test of fault-free operations for at least 25 years.

- Small: Real estate in telecommunications equipment is a valuable commodity. Light sources need to be small.

- Efficient: The light source must be able to convert the electrical signal into light efficiently without generating too much heat.

Two types of light sources fulfill all of these requirements: light-emitting diodes (LEDs) and laser diodes.

Transmission Through Air

Free-space optics, as the name implies, means the transmission of optical signals through free space or air. Such propagation of optical capacity through air requires the use of light. Light sources can be either LEDs or lasers (*light amplification by stimulated emission of radiation*). FSO is a simple concept that is similar to optical transmission using fiber-optic cables. The only difference is the medium. Interestingly enough, light travels faster through air (approximately 300,000 km/s) than it does through glass (approximately 200,000 km/s), so free-space optical communications could be classified as *optical communications at the speed of light.*

Environmental Challenges to Transmission Through the Air

Whereas fiber-optic cable is a predictable medium, free space, as an open medium, is less predictable (atmospheric attenuation is one example). Because of this unpredictability, it is more difficult to control the transmission of optics through free space. This unpredictability affects the system availability and maximum design capacities. FSO is also a line-of-sight technology, which means that the points that interconnect have to be able to see each other without anything in between. The main issues creating potential compromise of a link include the following:

- Fog: The major challenger to free-space optical communications is fog. To further qualify, it is dense fog that affects FSO connectivity. Fog is water vapor in the form of water droplets that are only a few hundred microns in diameter. These droplets are able to modify light characteristics or completely hinder the passage of light through them through a combination of absorption, scattering, and reflection.

- Absorption: Absorption occurs when suspended water molecules in the terrestrial atmosphere extinguish photons, causing a decrease in the power density of the beam (attenuation) and directly affecting the availability of a system. Absorption occurs more readily at some wavelengths than others.

- Scattering: Unlike absorption, scattering results in no energy loss, only directional redistribution of energy (multipath effects) that can cause a significant reduction in beam intensity, particularly for longer link distances. Three main types of scattering exist: Rayleigh, Mie, and nonselective scattering. Mie scattering, a scattering mechanism that becomes important when the particle size and wavelength are similar, is the main attenuation process to impact FSO system performance.

- Physical obstructions: Birds can temporarily block the beam, but this tends to cause only short interruptions, and transmissions are easily resumed. Multibeam systems address this issue.

- Building sway: The movement of buildings can upset receiver and transmitter alignment. You can overcome this issue in several ways, which will be discussed in subsequent chapters.

- Scintillation: Heated air rising from the ground creates temperature variations among different air pockets. This can cause fluctuations in signal amplitude, which lead to "image dancing" at the receiver end. The most familiar effects of scintillation in the atmosphere are the twinkling of stars and the shimmering of the horizon on a hot day.

These challenges will be discussed in more detail in the following chapters. In the end, the benefits outweigh the limitations.

Other Points of Comparison

Whereas it takes months—if not years—to enable fiber-optic communications, free-space optical communications can be implemented in a matter of weeks or even days at a fraction of the cost. As mentioned earlier, fiber deployments are a sunk infrastructure, which is lost when the customer leaves the building or decides to cancel the service. In contrast, FSO is a redeployable platform, thereby proposing a zero sunk cost model. Furthermore, because of FSO's flexibility and ease of deployment in multiple architectures, it offers an economic advantage over fiber optics. Another important aspect to take note is the environmental benefits of free-space optics. Fiber requires digging of trenches, which may cause pollution, cutting of trees, and destruction of historical landmarks. FSO does not; therefore, it is friendly to the environment.

The Role of FSO in the Network

Communications is the act of exchanging information among two or more individuals; a network is the physical infrastructure that enables this exchange to take place across space and time. When you make a telephone call from New York City to San Francisco, you are projecting your voice over 3,000 miles of physical space; when you leave a voicemail on an answering machine, you are communicating your thoughts across time. Because the distance over which two people can communicate is constrained by the limits of the human senses—how far your voice can carry or how far your eye can see—audio or visual information needs to be converted into formats suitable for transmission over distances that defy these limits. The physical infrastructure that enables the transmission of voice, video, and data comprise the communications network.

As mentioned previously, you can carry communications content between two points in space in multiple ways: RF Wireless transmission over the airwaves, electrical signal transmission over copper and coaxial cable, light transmission over fiber-optic cable, and now light over air using free-space optics. Optical networking involves the process of carrying communications content—voice, video, and data—over light signals, whether it be on fiber or air. Take the example of a telephone call. When you speak into the handset, your voice is converted by the telephone into an electrical signal. Optical networking involves taking this electrical signal as input and doing three things: 1) converting it into a stream of light pulses, 2) carrying the light signal over a fiber-optic cable or air to its destination, and 3) reconverting the light pulse back into its electrical format. Optical networking, after you sift through all the hype and the jargon, is nothing more than using light pulses to enable communication between two individuals. Conceptually, it is really that simple.

Summary

With only 5% of the buildings connected to fiber, the increase in high-bandwidth applications at the edge of the network, the lack of a high-speed infrastructure that connects the edge to the core, and increases in costs and time to lay fiber, the threat of the connectivity bottleneck becomes real. This threat not only impacts the end users, but also affects the service providers who face delays in laying fiber and building optical infrastructure. Time and cost are playing against these service providers, which results in incomplete networks, lack of revenue, and increased competition. What service providers need is a means to accelerate the completion of their optical networks and to access that traffic at the edge so that they can start to generate revenue quickly. Free-space optics provides them with such a solution and allows them to provide this optical connectivity not only cost effectively, but also quickly and reliably. Service providers can deploy FSO solutions where and when needed, as they see fit, in any topology, and with their exiting infrastructure—at a fraction of the cost—and generate revenue immediately. Such flexibility makes the FSO solution extremely attractive and can help service providers in truly solving this connectivity bottleneck.

Chapter 2, "Fundamentals of FSO Technology," examines some of the physical fundamentals of FSO so you can get a deeper understanding of its capabilities and limitations.

Sources

[1] Ryan Henkin Kent (www.rhk.com)

Fundamentals of FSO Technology

IN THIS CHAPTER

Before planning or installing an FSO system, it is important to understand the mechanics of the technology. You need to understand FSO basics from three angles: the physics behind free-space optics, the perspective of the system, and knowledge and understanding of transmission power. Together, these perspectives help explain the propagation of light through the air, which forms the basis of the free-space optics technology.

This chapter addresses the basics of free-space optical transmission technology. Its goal is to make you familiar with the general terminology used in FSO and provide an overview of system components, their functionality, and performance.

How FSO Works: An Overview

Free-space optics systems operate in the infrared (IR) spectral range. Commercially available FSO systems use wavelengths close to the visible spectrum around 850 and 1550 nm, which corresponds to frequencies around 200 THz. The 850 and 1550 nm wavelength ranges fall into two atmospheric windows (spectral regions that do not suffer much absorption from the surrounding atmosphere). Because these wavelengths are also used in fiber-optic communications, industry standard components on the transmission and receive side can be used.

The Federal Communications Commission (FCC) does not regulate frequency use above 300 GHz. Therefore, unlike most lower-frequency microwave systems, such as LMDS, FSO communication systems do not require operating licenses. This is true not only in North America, but also worldwide. Due to the proximity to the visible spectrum, the wavelengths in the near IR spectrum have nearly the same propagation properties as visible light.

In a basic point-to-point transmission system, an FSO transceiver (link head) is placed on either side of the transmission path. A main requirement for operating an FSO system is unobstructed line-of-sight between the two networking locations; FSO systems use light to communicate, and light cannot travel through solid obstacles such as walls or trees. A simple schematic of a free-space optics transmission system is shown in Figure 2.1.

The optical part of the transmitter involves a light source and a telescope assembly. The telescope can be designed by using either lenses or a parabolic mirror. The telescope narrows the beam and projects it toward the receiver. In practical applications, the beam divergence of the transmission beam varies between a few hundred microradiants and a few milliradiants. For example, for a 1-milliradiant beam divergence, the diameter of the beam at 1 km is 1 m. In practice, this is representative of moderate range FSO equipment.

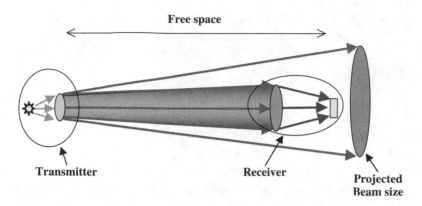

Free space

Transmitter

Receiver

Projected Beam size

FIGURE 2.1

Schematic of a free-space optical transmission system.

The transmitted light is picked up at the receiver side by using a lens or a mirror. Subsequently, the received light is focused on a photo detector. For all practical purposes, the projected beam size at the receiving end is much larger than the size of the receiving optics. Therefore, part of the transmitted light is lost during the transmission process. Depending on the actual beam divergence, the projected beam size can be several meters, whereas the typical diameter of the receiving telescope is more likely to be 8–20 cm. This phenomenon is called *geometrical path loss*. The use of a narrower beam decreases the amount of geometrical path loss. However, narrow beams require a very stable mounting platform or a more sophisticated active beam-tracking system.

An FSO system can operate in full duplex operation. This means that information can be received and transmitted in parallel and at the same time. Therefore, each FSO link head typically includes a transceiver capable of full duplex operation.

In a digital transmission system, the transmitter is modulated by an electric input signal that carries the actual network traffic. This is similar to the operation of a fiber-optic transmission line. During the electro-optic (E-O) conversion process, information is converted from the electrical into the optical domain. This simple conversion process allows for keeping the transmission path independent of the transported networking protocol. In other words, the basic FSO transmission system can operate as a physical layer one connection between networking locations. On the receiver side, a telescope picks up the modulated light signal and the receiver converts the optical bit stream back into an electrical signal.

In the following sections, we will discuss the various components and subassemblies of an FSO transmission system in more detail.

Transmitters

In modern FSO systems, a variety of light sources are used for the transmission of optical data. We will focus on semiconductor-based transmission sources because semiconductor lasers currently are the primary transmission media in commercial FSO systems. The main differences between these transmission sources are wavelength, power, and modulation speed. The price for a high-performance transmitter can vary from tens of dollars to thousands of dollars. The use of a specific transmission source is dictated by the specific target application.

Light-Emitting Diodes (LED)

Light-emitting diodes (LEDs) are semiconductor light-emitting structures. Due to their relatively low transmission power, LEDs are typically used in applications over shorter distances with moderate bandwidth requirements up to 155 Mbps. Depending on the material system, LEDs can operate in different wavelength ranges. When compared to narrowband (or single wavelength) laser transmission sources, LEDs have a much broader spectral range of operation. The typical full width half maximum (FWHM) emission spectrum varies between 20 and 100 nm around the designed center wavelength of operation. Infrared LEDs are also used as transmission sources in household remote controls. Advantages of LED sources include their extremely long life and low cost.

LED Operation and Characteristics

An LED is a semiconductor pn junction. A *pn junction* is a component that emits light when an external forward-bias voltage is applied. Figure 2.2 illustrates the circuit symbol, the junction, and the relevant energy band structure. The energy band model can be used to describe the operation of an LED.

The band energy model illustrates the two allowable energy bands. A forbidden region commonly referred to as *bandgap* separates these bands. W_g (band gap energy) is the energy width of the bandgap. In the upper energy level, which is called the *conduction band*, electrons that are not bound to an atom can move freely. In the lower band, which is called the *valence band*, unbound "holes" can move freely. These "holes" exist at locations where electrons left a neutral atom; consequently, a hole leaves a net positive charge. When an electron recombines with a hole, energy is released and the atom returns to a neutral state. Whereas an n-type semiconductor has an additional supply of free electrons, the p-type material has a number of free holes.

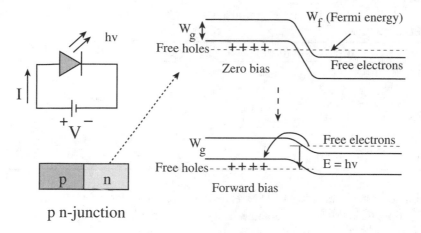

FIGURE 2.2

LED energy band gap model.

When an n- and a p-type material are brought together, the electrons and the holes recombine in the interface region. However, during this process, a barrier (neutral region) is generated and neither the electrons nor the holes have enough energy to cross this barrier. With zero bias voltage applied to the structure, the charge movement stops and no further recombination takes place. However, when a forward bias voltage is applied, the barrier decreases and the potential energy of the free electrons in the n-type material increases. In other words, the potential energy level of the n- side is raised, as can be seen in Figure 2.3. The forward bias voltage provides the electrons and holes with sufficient energy to move into the barrier region. When an electron meets a hole, the electron "falls" into the valence band and recombines with a hole. During this process, energy is released in the form of a photon. The wavelength of the light emitted during this process depends on the energy band gap width W_g, as shown in the following equation.

$$\lambda = \frac{h \cdot c}{W_g} \quad \text{or} \quad \lambda = \frac{1.24}{W_g}$$

The factor 1.24 provides the wavelength in micrometers when the bandgap energy is given in electron volts. Table 2.1 shows a listing of semiconductor material systems and the relationship between band gap energy and emission wavelength. For free-space optical applications, the Gallium Arsenide (GaAs) and Aluminum Gallium Arsenide (AlGaAs) material systems are of interest because the emission wavelengths fall into the lower wavelength atmospheric window around 850 nm. In the 850 nm wavelength range, the typical full width half maximum (FWHM) spectral width of an LED is 20–50 nm.

TABLE 2.1 Relationship Among Material System, Wavelength, and Band Gap Energy for Typical LED Structures

Material	Wavelength Range (um)	Bandgap Energy (eV)
GaInP	0.64–0.68	1.82–1.94
GaAs		0.9–1.4
AlGaAs	0.8–0.9	1.4–1.55
InGaAs	1.0–1.3	0.95–1.24
InGaAsP	0.9–1.7	0.73–1.35

The modulation bandwidth of an LED is related to the carrier lifetime τ, where τ is defined as the average time for carriers to recombine. The electrical modulation speed must be lower than the carrier lifetime. The electrical (3dB) bandwidth is given by:

$$f_{3dB} = \frac{1}{2\pi\tau}$$

LEDs typically operate at a modulation bandwidth between 1 MHz and 100 MHz. LEDs that can be used in applications that require a higher modulation bandwidth are not capable of emitting high optical power levels. A 1 mW LED is already considered to be high power at higher modulation speed.

Over time, the light output of an LED decreases for a given value of the driving current. However, the lifetime of LEDs (the length of time until the power is reduced to half of the original value) can be as high as 10^5 hours. This corresponds to about 11 years.

Some diodes tolerate temperature between –65 and +125 degrees Celsius. However, the output power decreases at higher temperatures.

With respect to light emission, LEDs are one of two types: surface-emitting LEDs or edge-emitting LEDs. Whereas surface-emitting diodes have a symmetric Lambertian radiation profile (a large beam divergence, and a radiation pattern that approximates a sphere), edge-emitting diodes have an asymmetrical elliptical radiation profile. LEDs are commercially available in a variety of packages: TO-18 or TO-46. Some packages include micro lenses to improve the quality of the output beam or to improve the coupling efficiency of the light into an optical fiber.

Laser Principles

A laser is similar in function to an LED, but somewhat different both in how it functions and in its characteristics. The word *laser* stands for Light Amplification by Stimulated Emission of Radiation. The idea of stimulated emission of radiation originated with Albert Einstein around

1916. Until that time, physicists had believed that a photon could interact with an atom only in two ways: The photon could be absorbed and raise the atom to a higher energy level, or the photon could be emitted by the atom when it dropped into a lower energy level.

Einstein proposed a third way of interaction: A photon with energy corresponding to that of an energy-level transition could stimulate an atom in the upper level to drop to the lower-level energy state, in the process stimulating the emission of another photon with the same energy as the first. Stimulated emission is unlikely to be seen in the thermodynamic equilibrium because more atoms are in the lower energy state than in the higher ones. Therefore, a photon is more likely to encounter an atom in a lower energy level and be absorbed than encounter an atom in a higher energy level and perform a stimulated emission process.

The first evidence of stimulated emission was reported in 1928 and it took another two decades before stimulated emission was more than a sophisticated laboratory experiment. The first demonstration of a solid state ruby laser was performed in 1960 at IBM. In 1961, a Bell Labs helium-neon laser demonstration took place. The race for a commercial version of a laser started shortly thereafter. The first semiconductor laser operation was demonstrated simultaneously in 1962 by three groups: General Electric, IBM, and MIT. Since then, the operation characteristics and performance of laser systems has continuously improved. However, the basic underlying concept of Einstein's stimulated emission stayed the same. The following discussion provides a short introduction to illustrate stimulated emission.

Figure 2.3 shows the typical energy diagram (term scheme) of an atom. An electron can be moved into a higher energy level by energy provided from the outside. As a basic rule, not all transitions are allowed, and the time that an electron stays in a higher energy state before it drops to a lower energy level varies. When the electron drops from a higher to a lower level, energy is released. A radiative transition that involves the emission of a photon in the visible or infrared spectrum requires a certain amount of energy difference between both energy levels.

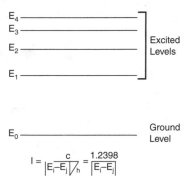

$$ l = \frac{c}{|E_i - E_j| / h} = \frac{1.2398}{|E_i - E_j|} $$

FIGURE 2.3
Energy level diagram.

Notice that the formula shown in Figure 2.3 is similar to the equation in the discussion of LED operation.

For ease of understanding, we will describe laser operation by using only two energy levels. Figure 2.4 illustrates the different methods of photon interaction.

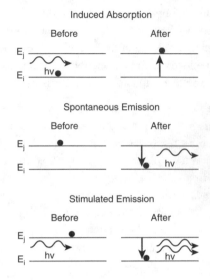

FIGURE 2.4
Understanding laser operation.

There are three possibilities:

- Induced absorption: An incoming photon whose wavelength matches the difference between the energy levels E_j and E_i can be absorbed by an atom that is in the lower energy state. After this interaction process, the photon disappears, but its energy is used to raise the atom to an upper energy level.

- Spontaneous emission: An atom in the upper energy level can spontaneously drop to the lower level. The energy that is released during this transition takes the form of an emitted photon. The wavelength of the photon corresponds to the energy difference between the energy states E_j and E_i. This resembles the process of electron-hole recombination that we discussed in the previous section, which resulted in the emission of a photon in the LED structure. Gas-filled fluorescent lights operate through spontaneous emission.

- Stimulated emission: An atom in the upper level can drop to the lower level, emitting a photon with a wavelength corresponding to the energy difference of the transition process. The actual emission process is induced by an incoming photon whose wavelength matches the energy transition level of the atom. The stimulated photon will be emitted in phase with the stimulating photon, which continues to propagate.

When these three processes take place in a media such as a solid-state material or gas-filled tube, many atoms are involved. If more atoms are in the ground state (or lower excited level) than in the upper one, the number of photons entering the material will decrease due to absorption. However, if the number of photons in the upper level exceeds the number of photons in the lower level, a condition called *population inversion* is created. Laser operation requires the state of population inversion because under these circumstances, the number of photons increases as they propagate through the media due to the fact that more photons will encounter upper-level atoms than will meet lower-level atoms. Keep in mind that upper-level atoms cause the generation of additional photons, whereas lower-level atoms would absorb photons. A medium with population inversion has gain and has the characteristics of an amplifier.

A laser is a high-frequency generator, or oscillator. To force the system to oscillate, it needs amplification, feedback, and a tuning mechanism that establishes the oscillation frequency. In a radio-frequency system, such feedback can be provided by filtering the output signal with a frequency filter, connecting the output signal back to the input, and electronically amplifying the signal before it is coupled back into the input stage. In the case of a laser, the medium provides the amplification. Therefore, a medium capable of laser operation is often referred to as active media.

The medium provides amplification through its characteristic energy levels and transitions between levels. At the same time, the medium determines its own frequency. In a laser system, mirrors provide the feedback mechanism. These mirrors form what is commonly referred to as laser cavity. Photons bounce off the mirrors and return through the medium for further amplification. At least one of the mirrors is partially transmitting light outside the cavity. However, most of the light is reflected back to establish the state of population inversion inside the cavity.

Laser oscillation occurs when the gain exceeds all losses in the laser system. These losses typically include absorption, scattering, and the extraction of laser power at the mirrors.

Similar to LEDs, lasers are electrically pumped. As long as the voltage is low, the gain is less than the loss and the output power is zero. When the voltage is increased slightly, a small amount of stimulated emission will occur. However, the power will be small, the output will not be coherent, and the spectral width will be large. At that point, the operation of the laser device is similar to an LED.

When the power is increased, more atoms are raised to the upper level and the internal gain increases. At a certain threshold voltage, the gain equals loss and oscillations start. Further increasing the voltage drastically increases the output power, and the light emission becomes coherent. At that point, the spectral width of the output radiation is extremely narrow.

The actual oscillation frequency can be determined by the geometry of the laser cavity. Only the laser modes that fall into the gain band of the laser and that fulfill the resonance conditions of the laser cavity can perform laser operation.

Figure 2.5 shows the relationship between input current and output power for a semiconductor laser diode system. Before the laser threshold is reached, the output power is low. At that point, the laser behaves similar to an LED. When the threshold is reached, a small variation of the input current causes a drastic increase of the output power level.

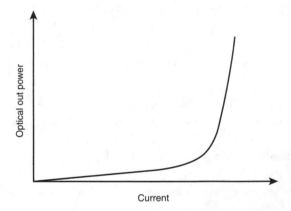

FIGURE 2.5
Relationship between input current and output power for a semiconductor laser.

A more detailed discussion of this topic is beyond the scope of this book. Details regarding the selection criteria can be found in any standard textbook covering the principle of laser operation. Within the next few sections, we will describe the operation of some specific laser systems that are used in free-space optics systems.

Laser Diodes

The entire commercial free-space optics industry is focused on using semiconductor lasers because of their relatively small size, high power, and cost efficiency. Most of these lasers are also used in fiber optics; therefore, availability is not a problem.

From the semiconductor design point of view, two different laser structures are available: edge-emitting lasers and surface-emitting lasers. With an edge emitter, the light leaves the structure through a small window of the active layer and parallel to the layer structure. Surface emitters radiate through a small window perpendicular to the layer structure. These variations also exist for LED. Figure 2.6 illustrates these two designs. Both have certain advantages and disadvantages when factors such as power output levels, beam quality, or mass production are taken into consideration.

Edge emitters can produce high power. More than 100 milliwatts at modulation speeds higher than 1 GHz are commercially available in the 850 nm wavelength range. The beam profile of edge-emitting diodes is not symmetrical. A typical value for this elliptical radiation output pattern is 20×35 degrees. This specific feature can cause a problem when the output power has to be coupled efficiently into a fiber and external optics such as cylindrical lenses are used to increase the coupling efficiency. Surface-emitting diodes typically produce less power output. However, the beam pattern is close to being symmetrical or round. A typical value for the beam divergence angle is 12 degrees. This feature is beneficial for coupling light into a (round) optical fiber.

Besides discussing basic designs of semiconductor lasers, we will also provide information regarding WDM laser sources and look into Erbium Doped Fiber Amplifiers/lasers that have been discussed recently for use in FSO systems.

Materials and Wavelengths

Similar to the gas used in a gas laser system, the specific material system (active media) determines the wavelength of operation in a semiconductor laser. Finding the correct material system was one of the major problems in the early days of semiconductor research and production. When semiconductor fabrication technologies such as molecular beam epitaxy (MBE) and chemical vapor deposition (CVD) were in their infancy, it was difficult to grow complex multi-layer semiconductor structures. As a result, the current that had to be provided to the semiconductor system before laser operation started was very high. This drastically impacted the lifetime of these lasers. The internal light field could not be confined well, and the quality of the mirrors was low, causing high loss coefficients within the resonant cavity structure.

Most of these problems have been resolved. The lifetime of semiconductor lasers has been increased dramatically when compared to the initial laser system designs. In most cases, semiconductor lasers are the preferred choice for companies that need a high power and coherent light source in their system design.

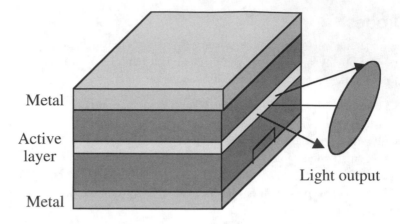

Metal

Active
layer

Metal

Light output

Edge-Emitting Diode

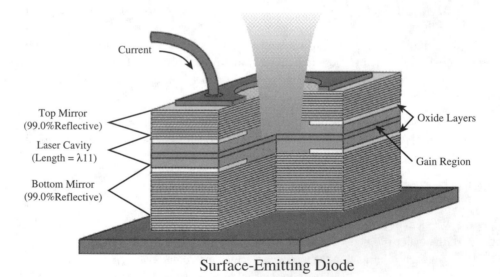

Current

Top Mirror
(99.0%Reflective)

Laser Cavity
(Length = $\lambda 11$)

Bottom Mirror
(99.0%Reflective)

Oxide Layers

Gain Region

Surface-Emitting Diode

FIGURE 2.6

Edge- and surface-emitting diode structures.

Many semiconductor laser systems can be custom designed. Because the bandgap of a semi-conductor depends on the crystalline structure and chemical deposition of the material, diode lasers can be tailored to operate at a specific wavelength by changing the composition of the material system. This is especially an advantage when the application calls for narrowly spaced wavelengths, such as in fiber WDM systems.

Table 2.2 gives an overview of material systems and corresponding wavelength ranges that are relevant in free-space optical communication. Output power levels that can be achieved by various semiconductor lasers vary between a few milliwatts to several hundred milliwatts. The actual power levels required for a specific application depends on factors such as bandwidth, distance, and so on. This topic will be addressed in the section on link margin analysis later in this chapter.

TABLE 2.2   Relationship Between Material Systems and Wavelengths

Compound	Wavelength, nm	Remarks
Ga(1–x)Al(x)As	620–895	X= 0–0.45. Short lifetimes for wavelengths < 720 nm.
GaAs	904	
In(1–x)Ga(x)As(y) P(1–y)	1100 -1650	InP substrate.
In(0.58)Ga(0.42)As(0.9)P(0.1)	1550	Major fiber communication wavelength.
InGaAsSb*	1700–4400	Possible range, developmental, on GaSb substrate.
PbEuSeTe*	3300–5800	Cryogenic.

Currently, these wavelength are not used in FSO systems.

Laser Design

There are a myriad of laser types and laser cavity configurations. The purpose of the cavity is to confine light and create the resonant condition for specific laser wavelengths.

One diode laser cavity design seen often is the Fabry-Perot cavity. In it, the light is confined in the active layer by using semiconductor materials (that is, aluminum, gallium, and Arsenide) that are somewhat similar to each other but have different refractive indices. Using AlGaAs material, the refractive indices can be changed by using a slightly different composition of the materials. The reflectivity coefficients are high so that one mirror nearly completely reflects the light, whereas the other mirror is slightly transparent. In this case, a small portion of the light escapes the cavity; this is the laser's output.

An enhancement to laser performance is seen in the distributed feedback (DFB) laser. It emits a narrow spectrum of light, nearly a single wavelength (< 0.1 nm available). DFB lasers are more costly (up to 1,000 times the cost of a basic Fabry-Perot laser), can require thermoelectric cooling, but can provide performance benefits.

The Distributed Bragg Reflector (DBR) laser can be used to provide a "tunable" laser to output wavelengths. Although it is expensive, the DBR laser could be used to adjust laser wavelength for specific atmospheric conditions.

A popular choice in FSO equipment is the Vertical Cavity Surface Emitting Laser, or VCSEL (pronounced *vixel*). Rather than emitting light out the side of the chip, the VCSEL laser emits light up out of the wafer, perpendicular to the surface. VCSELs have the advantages of low power consumption, low heat generation, easily coupled elliptical output beam at the facet, low cost, and high bandwidth (up to 5 GHz).

WDM Laser Sources

Laser sources that are suitable for wavelength division multiplexing (splitting up the light to achieve multiple channels for increased throughput) are single mode and narrow in spectral bandwidth. For system interoperability, ITU established a standard around WDM operation with the intention to unify the standard on an international level. The operating center frequency (wavelength) of channels must be the same on the transmitting and receiving side. As of October 1998, the ITU-T G.692 standard is in place that recommends 81 channels starting from 196.10 THz and decrementing by 50 GHz (0.39 nm). This wavelength band overlaps with the amplification band of Erbium Doped Fiber Amplifiers, or EDFAs.

The first frequency in this so-called C-band starts at 196.10 THz or 1528.77 nm, and the last frequency is 192.10 THz or 1560.61 nm. Another way that was initially chosen to express this standard was the formula

$$\lambda = 193.1 \text{ THz } +/- \ m * 100 \text{ GHz}$$

with m being an integer.

The 100 GHz spacing corresponds to a wavelength difference close to 0.8 nm. This standard specifies the wavelength grid for what is also known as dense wavelength division multiplexing, or DWDM. Initially, many systems operated at a 200 or 400 GHz spacing, but with improvements on the manufacturing side and requirements for higher capacity (channel counts), operation was pushed more toward the 100 and even 50 GHz spacing. Coarse WDM operation using just one wavelength in the 1300 nm and one wavelength in the 1550 nm wavelength band was used before, but the real DWDM "revolution" in long-haul communication systems really began after this standard started to unfold in the mid 1990s.

Today's sophisticated WDM laser sources follow the wavelength grid specifications. Most lasers are of DFB or DBR type and incorporate electronics and a temperature control mechanism such as a TE cooler to stabilize or fine-tune the wavelength according to the DWDM specification. Due to the tight wavelength specifications, the production process of these devices is controlled extremely precisely. One of the obvious reasons is that the resulting laser wavelength is controlled by the geometric parameters of the cavity. The production yield of laser sources suitable for DWDM operation is still low; therefore, these sources are still quite expensive when compared to standard laser sources that operate at an unspecified wavelength.

Similar to fiber-optic systems, WDM operation in FSO systems is an appealing approach because it allows for increasing the transmission capacity by adding wavelengths. Because the 1550 nm band falls into a transmission window of the atmosphere, standard lasers and components that are already used in fiber-optic systems can be used in WDM FSO systems, too. The perfect overlap of the low attenuation window around 1550 nm in optical fibers and the atmosphere allows for building all optical transmission systems incorporating both fiber and FSO transmission in a transparent way. In laboratory trials, FSO systems operating with up to 40 separate wavelengths have been demonstrated successfully.

Wavelength dispersion, or nonlinear effects such as four-wave-mixing that somewhat limit the potential of DWDM operation in optical fiber, are not a major concern in FSO DWDM systems. Therefore, DWDM operation seems to be even better suited for FSO systems.

Erbium Doped Amplifier Sources

In current FSO systems, the signal undergoes many stages of optical-electrical-optical (O-E-O) conversion. This regeneration and amplification process increases the complexity of FSO systems because it involves several data-rate-dependent pieces of electronics. A much more straightforward way to amplify the optical signal involves the use of an optical amplifier such as an Erbium Doped Fiber Amplifier (EDFA). EDFAs are used to directly boost the optical power output of a laser source. In an EDFA, the signal remains in the optical domain throughout, no O-E-O conversion takes place, and consequently, no data-rate or protocol-dependent electronics are required. Figure 2.7 illustrates the difference between a more complex O-E-O conversion process and an "inline" amplification process using an EDFA.

An EDFA is basically a fiber segment heavily doped with erbium atoms. The erbium atoms can be excited into a higher energy state by a number of wavelengths, including 532, 667, 800, 980, and 1480 nm. Figure 2.8 shows a simplified atomic term scheme of an erbium (Er^{3+}) atom and illustrates the pump process for the most commonly used pumping wavelength of 980 and 1480 nm.

When erbium ions are excited by a 980 nm source, the excited ions fall back after approximately 1 microsecond in a lower metastable energy state that has a spontaneous lifetime of about 14 ms. The transition from the low lifetime to the high lifetime state is a nonradiative transition process that does not generate photon emission. The Er ion can end up in the same low energy state by directly using a 980 nm pump source. The typical pump power varies between 100 mW up to around 1 watt. These pump levels create the population inversion that is required for a stimulated emission process.

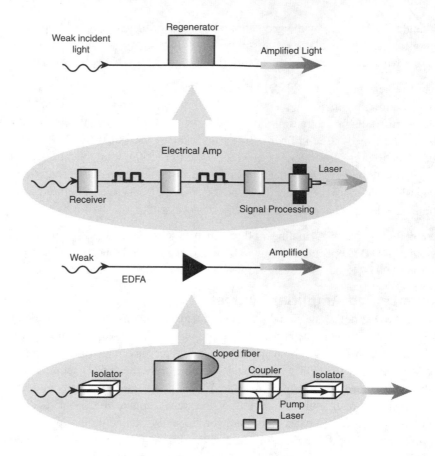

FIGURE 2.7

Regenerator complexity versus EDFA simplicity.

If the atoms are triggered to release the energy by a photon in the 1.5 micron wavelength band traveling through the Er-doped and excited region of the fiber, they release the energy and fall back into the energy ground state. If population inversion is present, this process is a stimulated emission process that amplifies the light field.

Depending on the wavelength of the incoming photon, this amplification process can take place over the whole gain band of the Er^{3+} transition band. In an EDFA amplifier, the wavelength band between 1520 and 1620 nm can be amplified. However, the gain factor is not always the same; therefore, special provisions are necessary to ensure equal amplification of wavelength if this is required within the overall system design.

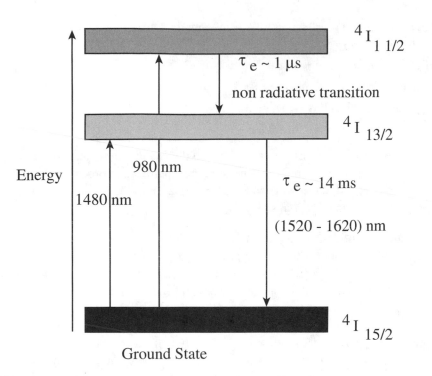

FIGURE 2.8

Energy level diagram and pumping of Er³⁺ ions.

In communication systems, the bit rate is high. The typical bit duration is less than 1 ns; consequently, it is short compared to the typical spontaneous emission lifetime of the excited atom, which is around 14 ms. However, if no light signal is present, atoms release their energy by spontaneous emission. Because these photons will be amplified while traveling through the erbium doped fiber, this emission is called *amplified spontaneous emission* (ASE). ASE adds to the noise figure of the amplifier and is not desirable when the light is detected at the receiver side.

In today's optical communication systems, EDFA technology is the preferred amplifier technology for all-optical amplification of light in the 1550 nm wavelength range. The following bullets highlight the benefits of EDFA technology:

- High pump efficiency (50%)
- Direct and simultaneous all-optical amplification of a wide wavelength band in the 1550 nm wavelength region that matches the low attenuation band of optical fiber and FSO systems

- High power output (as high as +37 dBm) with a relatively flat gain (>20 dB) for DWDM applications
- High saturation output power
- Relatively low noise figure (<5 dBm)
- Optically transparent to modulation format
- Polarization-independent operation

Some of the disadvantages of EDFAs are related to the fact that they cannot be easily integrated with other semiconductor devices because the amplification process takes place in a longer piece of fiber and not in semiconductor material. The ASE factor limits the number of amplifiers that can be connected in series.

Laser Diode Selection Criteria for FSO

The selection of a laser source for FSO applications depends on various factors. It is important that the transmission wavelength is correlated with one of the atmospheric windows. As noted earlier, good atmospheric windows are around 850 nm and 1550 nm in the shorter IR wavelength range. In the longer IR spectral range, some wavelength windows are present between 3–5 micrometers (especially 3.5–3.6 micrometers) and 8–14 micrometers. However, the availability of suitable light sources in these longer wavelength ranges is pretty limited at the present moment. In addition, most sources need low temperature cooling, which limits their use in commercial telecommunication applications. Other factors that impact the use of a specific light source include the following:

- Price and availability of commercial components
- Transmission power
- Lifetime
- Modulation capabilities
- Eye safety
- Physical dimensions
- Compatibility with other transmission media such as fiber

Receivers

Besides transmission sources, light detectors are important building blocks within the FSO system design. Light receivers detect light by using different physical phenomena. Similar to laser sources, most detectors used in commercial FSO systems are semiconductor based. Depending on the specific material system, they can operate in different wavelength ranges. This section describes some of the basic considerations of receiver configurations.

Principles of Light Detection

In modern high-speed communication applications, two important physical mechanisms are used to detect light signals: the external and the internal photoelectric effect. Both of them convert the incoming photon energy into electrical energy. Vacuum diodes or photomultipliers are based on the external photoelectric effect, whereas semiconductor detectors such as PIN or Avalanche diodes use the internal photoelectric effect to detect photons. Some important factors—such as responsivity, spectral response, and rise time—will be reviewed next. They are useful when comparing detector capabilities.

The responsivity ρ of a detector defines the relationship between the output current of a photodetector to its optical power input.

$$\rho = i/P$$

In this case, Ampere/Watt (A/W) gives the physical value for the responsivity of a detector.

The spectral response is related to the wavelength sensitivity region of a specific detector. It provides a figure related to the amount of current produced at a specific wavelength, assuming that all wavelengths provide the same amount of light power.

The *rise time* is the time is takes for the detector to raise its output current from 10% to 90% of its final value, when a step-shaped light pulse is applied to the surface of the detector. The 3 dB modulation bandwidth of a detector is related to the rise time t_r:

$$f_{-3dB} = 0.35/t_r$$

For the intrinsic photoelectric process, the quantum efficiency η provides the number of electrons generated divided by the total number of photons:

$$\eta = \frac{Number of output electrons}{Number of input photons}$$

The responsivity ρ and the quantum efficiency η are connected through the following relationship:

$$\rho = \frac{e\eta}{hv}$$

in which e is the electron charge, h is the Planck constant, and v is the light frequency.

In terms of photocurrent, this formula can be rewritten as follows:

$$i = \frac{\eta e\lambda}{hc}$$

in which λ is the wavelength of the incident photon, and c is the vacuum speed of light.

Semiconductor Photodiode

Semiconductor photodiodes are small, fast, and sensitive and provide many different wavelength bands that are relevant for FSO systems. The simplest form of semiconductor diode is a *pn* diode. The basic detection mechanism of the junction detector is simple. As can be seen in Figure 2.2, the potential energy barrier between the p and n region of a semiconductor material increases when a reverse bias voltage is applied. The free electrons in the n region and the free holes in the p region do not have enough energy to climb the potential barrier; therefore, no current flow can be observed in an outside circuitry that connects both materials. The barrier region is called the *junction*. Because of the nonexistence of carrier in the junction, the junction is considered depleted. Most of the voltage drop can be measured across the junction where the electric field across the p and n layer is relatively weak.

When a photon enters the structure through the p layer and is absorbed inside the junction, it can raise the energy of an electron across the barriers and consequently move the electron from the valence into the conduction band. In other words, the photon created a free electron or an electron-hole pair. Driven by the electric fields, the hole and the electron start to move in opposite directions through the respective p and n layers. These moving charges cause a current to flow in an outside circuit. They can be measured as a voltage drop across a resistor located in the outside circuit.

When a photon is already absorbed either inside the p or n layer, it creates an electron-hole pair, but due to the low electric-field strength in these regions, these charge carriers will only slowly diffuse, and most of them will recombine before reaching the junction area. These charges produce a negligible current, but reduce the detector sensitivity. This is the main reason why pn detectors are inefficient. This carrier diffusion issue is also the reason pn detectors are relatively slow in response time. The typical response time of a pn detector is in the microsecond range, thus making them unsuitable for high-speed light detection.

PIN Diodes

PIN diodes solve the problem of low responsivity and slow rise time in semiconductor structures. They are the most commonly used semiconductor detectors in FSO equipment.

As illustrated in Figure 2.9, the PIN diode has a wide intrinsic semiconductor layer separating the p- and n- layers. The intrinsic layer has no free charges, so its resistance is high. Therefore, most of the external voltage drop appears across the intrinsic layer, and the electric field forces within the layer are high.

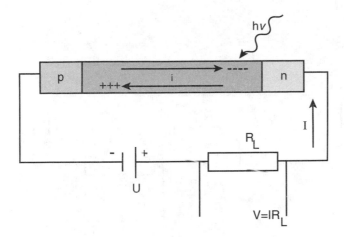

FIGURE 2.9
Schematic of a PIN diode structure.

Due to the wide intrinsic layer design, most of the photons will be absorbed within the intrinsic region rather than within the p- and n- layers at the outside of the structure. This drastically improves the responsivity and rise time of PIN diodes when compared to pn diodes. To create an electron-hole pair inside the intrinsic layer, the photon needs a minimum amount of energy to lift the electron across the bandgap. Because the energy of the photon is related to its wavelengths, the cut-off wavelength λ_c for a specific detector material is given by the following:

$$\lambda_c = \frac{1.24}{W_g}$$

with λ given μm, and W_g in eV. This is the same equation we used before for LED or laser emitters.

Avalanche Photodiodes (APD)

An avalanche photodiode, or APD, is a semiconductor detector that has internal gain. This increases the responsivity when compared to pn or PIN detectors. Internal gain yields better signal-to-noise ratios when compared to using an external amplification stage, such as an external transistor circuitry. Figure 2.10 shows the internal structure of an APD. The specific type of APD shown is called *reach-through APD*.

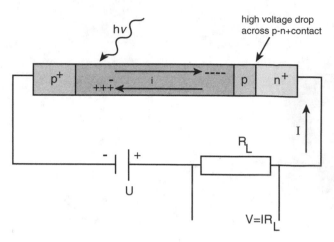

FIGURE 2.10

Schematics of a reach-through APD structure.

Avalanche current multiplication can be described in the following way: A photon absorbed in the depletion region π generates an electron-hole pair. Similar to the PIN structure, this depletion region is an intrinsic region that contains only a few free-charge carriers; therefore, it is highly resistant. Because the voltage drop across the depletion region is high, the charge carriers created by the incoming photon are accelerated in the electric field and gain kinetic energy. If the kinetic energy gained during this acceleration process is high enough to create another electron-hole pair during the collision with another atom, a secondary electron-hole pair is generated.

This process can take place repeatedly, multiplying the carriers through an avalanche-like process. To make the process effective, a large reverse bias voltage must be applied to the diode structure. In some instances, these voltages can be several hundred volts.

In the reach-through diode structure, the regions marked by n+ and p+ are heavily doped with charge carriers. Due to the high amount of carriers inside these layers, their resistance is low; consequently, a low voltage drop also is present across them. As mentioned before, the π region is only lightly doped (nearly intrinsic) and has a rather high resistance. As a result, the depletion region at the pn+ junction has "reached through" to the π layer. The voltage drop is mostly across the pn+ junction, where the resulting large electrical fields cause carrier multiplication when the electrons that are generated in the π layer enter this region. Holes that are generated at the same time in the π layer drift toward the opposite p+ layer. However, they do not reach the needed amount of kinetic energy to create additional electron-hole pairs. Therefore, they do not take part in the multiplication process, which is beneficial because structures that have only one carrier type (electrons or holes) involved in the multiplication process have superior noise properties.

The gain of an avalanche photodiode increases with increasing bias voltage according to the following formula:

$$M = \frac{1}{1 - \left(\dfrac{v_d}{V_{BR}}\right)^n}$$

in which V_{BR} is the reverse breakdown voltage of the diode, v_d is the reserved bias voltage, and n is an empirical factor that is more than 1. The breakdown voltages depend on the specific diode structures and can vary roughly between 20 and 500 volts.

The responsivity ρ for a gain-driven APD diode can be written as follows:

$$\rho = \frac{Me\lambda\eta}{hc}$$

in which η is the quantum efficiency with unity gain. Typical avalanche responsivities range from 20–80 A/W. This is considerably higher when compared to silicon PIN diodes that have typical responsivities between 0.5–0.7 A/W.

The responsivity ρ can be translated into a photo current i by using the following formula:

$$i = \frac{Me\lambda\eta P}{hc}$$

in which P is the optical input power.

Receiver Selection Criteria for FSO

Similar to light sources, the choice of a specific type of detector or detector material depends on the application. The sensitivity characteristic has to match the transmission wavelength of the transmitter.

For shorter wavelength applications in the 850 nm wavelength window, silicon detectors are certainly the best choice. PIN detectors are sufficient for applications over shorter distances and when the opposite transmitter can provide sufficient power. APDs are certainly much better for applications over longer distances. The higher sensitivity of the APD design provides additional link margin. However, APDs require a stable and high-bias voltage, and they are more expensive than PIN diodes. In addition, the typical dark current of APD diodes is higher when compared to PIN diodes. (*Dark current* refers to the external current that, under specified biasing conditions, flows in a photoconductive detector when there is no incident radiation.)

The silicon material system has a steep cut-off wavelength at 1.1 µm. Therefore, silicon cannot be used in applications with longer wavelengths. For the 1550 nm range, InGaAs is the material system of choice. The responsivity of the InGaAs material systems can be as high as 0.9 A/W around 1550 nm. InGaAs PIN diodes are widely available commercially. They have excellent modulation characteristics and can operate at high speeds (10 Gbps and higher).

Germanium has a wide spectral response and can operate in the short and longer wavelength atmospheric windows. However, germanium has very high values of dark current; therefore, it is not used often in FSO applications.

For potential applications in the 3–5 µm and 8–14 µm wavelength ranges, more exotic detector materials such as Mercury cadmium telluride (MCT) with spectral responses in these wavelength ranges can be used. However, similar to the transmission sources, these materials must be cooled at low temperatures.

Optical Subsystems

Optical subsystems play an important role in the overall FSO system design. Optical components are used on the transmission as well as on the receive side of an optical link. In modern FSO systems, different lens- and mirror-based designs are used. Whereas lenses are based on the physics of light refraction, mirrors are based on reflective properties of materials. The design chosen often depends on the performance requirements for the specific application and the price points. In this chapter, we review some of the fundamentals of classical optics. For a deeper understanding, you can review many good textbooks that cover specifics of optical design. The main goal of this chapter is to familiarize you with some specific optical designs used in FSO systems.

Classical Ray Lens Optics

In classical ray or geometrical optics, light is considered to consist of narrow beams. In this case, optics phenomena can be described by using a principle geometric approach. Geometrical optics uses a few simple rules:

- The relationship between the speed of light in a vacuum (*c*) and the actual speed of light in a media (*v*) other than vacuum is given by the following equation:

$$v = \frac{c}{n}$$

  in which $c = 3 \times 10^8$ m/s and n is the refractive index of the medium. Because the refractive index of air and other gases is close to unity, light is only slowed nominally in gases even under nonvacuum conditions. For silicon, for example, the refractive index is close to 3.5, and glass has a refractive index close to 1.5. However, different glass materials have slightly different refractive indices. This fact is used in fiber optics to provide waveguide structures (structures that have the ability to guide optical energy).

- Light rays are traveling in a straight direction unless they are deflected by a change in the media along the propagation path.

If light enters the boundary between two media with different refractive indices, the light rays are reflected back at an angle equal to the angle of incidence. These angles are measured with respect to the *normal* direction that is perpendicular to the boundary. Figure 2.11 illustrates this behavior.

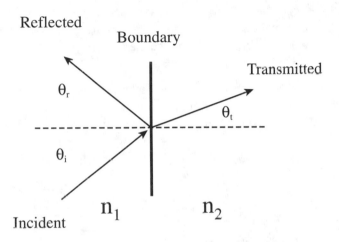

FIGURE 2.11
Behavior of incident, reflected, and transmitted light rays at the interface of two media with different refractive indices n_1 and n_2.

- The angle of the transmitted beam is related to the angle of the incident beam by Snell's Law:

$$\frac{\sin\theta_t}{\sin\theta_i} = \frac{n_1}{n_2}$$

Snell's Law has two important implications:

- The transmitted beam is bent toward the normal direction when the incident light enters from a material with a lower refractive index into a material with a higher refractive index.
- The transmitted beam is bent away from the normal direction when the incident light enters from a material with a higher refractive index into a material with a lower refractive index.

Basic Designs of Optical Lenses

A lens is a piece of glass or other transparent material that refracts light rays in such a way that they can form an image. Lenses can be envisioned as a series of tiny refracting prisms, and each of these prisms refracts light to produce its own image. When the prisms act together, they produce an image that can be focused at a single point.

Lenses can be distinguished from one another in terms of their shape and the materials from which they are made. The shape determines whether the lens is converging or diverging. The material has a refractive index that determines the refractive properties of the lens. The horizontal axis of a lens is known as the *principal axis*.

A converging (convex) lens directs incoming light inward toward the center axis of the beam path. Converging lenses are thicker across their middle and thinner at their upper and lower edges. When collimated (parallel) light rays enter a converging lens, the light is focused to a point. The point where the light converges is called the *focal point* and the distance between the lens and the focal point is called *focal length*.

A diverging (convex) lens directs incoming rays of light outward away from the axis of the beam path. Diverging lenses are thinner across their middle and thicker at their upper and lower edges. Figure 2.12 illustrates the behavior of converging and diverging lenses.

Converging Lenses   Diverging Lenses

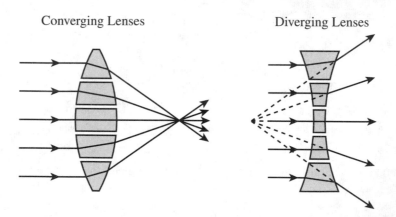

FIGURE 2.12

Converging and diverging lenses.

More specifically, the lenses in Figure 2.12 are double convex (converging) and double concave (diverging) lenses, respectively. Such lenses are symmetrical across both their horizontal and vertical axes.

Besides double convex and concave lenses, plano convex and concave lenses are often used in optical designs. These lenses have one plane, as well as one either convex- or concave-shaped surface.

Optical Designs for Free-Space Optics

A number of optical system designs are commonly used in FSO applications. These optical designs tend to be relatively simple to reduce costs and system complexity. Optics are used to transmit light from emitters with low divergence, collect light on detectors, and couple light into optical fiber. The optical system is needed to control light with a narrow bandwidth in the Near Infrared (NIR) region of the optical spectrum. Other requirements to consider include aperture diameter, $F_\#$, full field of view (FFOV), resolution, and overall length (OAL) of the optical assembly.

To optimize the system performance, match the optical system to the detector size and transmitter beam divergence. With fiber-coupled systems, the numerical aperture (NA) of the fiber is matched to the $F_\#$ of the optical system via $F_\# = 1/2(NA)$. Single mode fiber is generally $F/_5$ (NA = 0.1), and multimode fiber is $F/_{1.8}$ (NA = 0.275).

Tracking and Acquisition

Tracking and acquisition of laser beams has been one of the major topics in conferences covering various aspects of satellite-based laser communication systems. For communication between satellites or communication between satellites and ground-based laser terminals, the precise pointing of the laser beam is a major issue. Distances between these remote locations can be hundreds of kilometers, and the beam must be narrow (a few μrad) to transport as much light power as possible to the opposite receiver. A slight mispointing of such a narrow beam could cause a complete interruption of the communication link.

For space-based applications, not only automatic tracking but also automatic and remote acquisition of the remote communication side is important. No one on the satellite can align the laser link toward the opposite remote satellite. Numerous methods for coarse and fine tracking and automatic acquisition have been developed to accomplish this task. These methods include the use of servo motors, stepping motors, voice coils, mirrors, quad detectors, CCD arrays, and even liquid crystals and micro-electromechanical systems (MEMS). Generally, sophisticated tracking systems that have been developed specifically for outer space applications are not suitable for terrestrial applications due to high cost. We would like to refer the reader who is interested in learning more about this topic to study the SPIE proceedings on free-space laser communications technology. These proceedings can be ordered directly from the SPIE Web page (`http://www.spie.org`).

In terrestrial-based FSO applications, the distances involved between remote sides are much smaller but misaiming of the beam is still a significant issue. Imagine that your task is to focus the spot of light from a presentation pointer exactly on the dot of an "i" 100 feet away and hold it there. Not an easy task. Operating wider beams and accepting a higher geometrical path loss are acceptable tradeoffs for ease of deployment.

Wide Beam Transmission Systems

Wide beam transmission systems without a tracking feature are a cost-effective and reliable solution for operation at moderate speed and over moderate distances. The wider beam causes an increased loss of transmission power, and this must be reflected in the link budget calculation. When a circular beam is used, the total amount of power received for a given size of the receiver surface that is located at a fixed distance will increase by 6 dB when the diameter of the projected beam is cut in half. Reducing the diameter by a factor of two translates to decreasing the divergence angle by a factor of two.

Standard and commercially available FSO systems that use wide-angle transmission without tracking operate at divergence angles between 2–10 milliradians (mrad). A 2–10 mrad divergence angle roughly corresponds to a beam size diameter of 2 m and 10 m at a distance of 1 kilometer. Most FSO vendors with deployment in the field found out that these beam divergence angles provide sufficient mispointing angle margin to keep the beam on target. The key to minimization of mispointing difficulties is the selection of a stable mount and mount location when the system is installed. Mountings on masonry sidewalls are preferable to wood; sidewall corners of buildings are better than rooftops. Typical equipment mount rigidity specifications are < 1-3 mrad of allowable mispointing angle. However, if the system is installed on a high-rise that experiences a large amount of swap, an active tracking system is beneficial to counteract the mispointing of the beam at the receiver side. In addition, the installation of FSO systems on tall and unstable towers, telephone poles, and other kinds of unstable platforms is not recommended without using an active beam-tracking system.

Auto Tracking

Auto tracking is a feature in which the beam is automatically realigned toward the opposite receiver in case of building or tower sway. Auto trackers incorporate a mechanism that detects the position of the beam at the receiving side and a counter measure that controls and keeps the beam on the receiving detector. Many tracking systems use a beacon beam that is separate from the data-carrying beam to accomplish this task. In this case, it is important that the data and beacon beams are lined up in the same direction.

To distinguish the data and the beacon beam, two different wavelengths can be used. A simple *dichroic mirror* (a mirror used to reflect light selectively according to its wavelength) that reflects the beacon beam onto a position-sensitive detector that is transparent for the data signal wavelength is used. If the same wavelength is used for data transmission and tracking, a beam splitter can be incorporated into the optical path that reflects part of the incoming light onto the position-sensitive detector element. However, one disadvantage of this method is that a certain amount of the incoming light signal that carries the data traffic will be lost. Auto trackers very often use a close loop feedback control mechanism to keep the receive beam on target.

Gimbals

A *gimbal* is a device that often is used to support a link head and can be turned in different directions; it typically covers the motion in the vertical as well as in the horizontal direction. It can swing in two axes: up and down and side to side. You might be familiar or might have seen gyroscopes that are mounted in a gimbal arrangement.

Gimbals are useful for performing the automatic acquisition and tracking of the remote side. Important characteristics of a gimbal are shown in Table 2.3.

TABLE 2.3 Gimbal Characteristics

Characteristics	Typical Values
Vertical Field of Regard	+/- 20 degrees
Horizontal Field of Regard	+/- 25 degrees
Gimbal Jitter	< 5 radians rms
Slew Rate	20 radians/sec
Acceleration	Azimuth = 7 rad/sec^2 Elevation = 12 rad/sec^2)

Servo-Based Tracking Systems

Servomotors that drive a belt to rotate a gimbal tracker are often used in gimbal designs. Servomotors are robust, but they can have a high power consumption, which can cause a problem when systems are installed on rooftops.

Steering Mirror-Tracking Designs

The gimbal described previously actually moves the entire telescope during the tracking or acquisition process. This results in having to accelerate a physical mass of material of considerable weight, and there are certain limitation on how fast you can track a moving target. Therefore, these designs are typically used for slower-speed tracking.

A steering mirror is basically a mirror that is mounted onto a platform that can change the mirror direction. By using voice coils or actuators (such as piezo elements), this movement of the mirror can be three-dimensional. In tracking applications, this approach can be used to deflect a beam of incoming light in different directions, manipulating the driving voltage of the voice coils or the piezo elements, respectively.

The obvious advantage of this approach is that no huge masses are involved that have to be moved. The mirror is normally lightweight. This allows for fast tracking of the incoming light beam. Tracking speeds up to several hundred hertz are possible by using this procedure.

One drawback is certainly the inability to follow larger tracking angles. In space-based satellite applications, the large-angle gimbal tracker and the steering-mirror design are combined to ensure highest flexibility and fastest response time. In terrestrial FSO applications, the large angle-tracking capability over 20 degrees or more in both horizontal and vertical direction is not of great concern if the system is permanently installed to connect two buildings in a point-to-point scenario. In these cases, a steering mirror–based tracking system provides sufficient beam-tracking flexibility. This results, of course, in a high cost saving for the potential user of FSO systems.

Micro-Electromechanical Systems (MEMS)

Silicon micromachines are the mechanical analog of silicon electronic integrated circuits, and they are fabricated by similar methods. The application range for these kinds of devices range from data modulators, variable attenuators, optical switches, cross connects, and drop multiplexers to steering mirrors. For tracking applications, micromachined steering mirrors are of great interest. They are small, low on power consumption, and have a fast enough response time to counteract potential building or tower swap. Micromachined steering mirrors can be mass-produced and fully integrated with the receiver on a small footprint assembly. In conjunction with a digital signal processor that runs the tracking algorithm in the background, micromachined mirrors have a high potential to become a powerful and cost-effective approach in terrestrial FSO tracking.

Based on their analog micromirror technology, Texas Instruments (TI) (http://www.ti.com) recently introduced a MEMS steering device that can be incorporated into FSO systems. This device has the following technical data according to the TI Web page:

Material: Single crystal silicon

Mirror area: 3.2 mm×3.6 mm

Die size: 7 mm×9 mm

Mirror curvature: > 40 meters

Mirror surface to pivot point: 50 mm

Reflectivity: > 97% (840 and 1550 nm)

Range of motion: 2-axis

Beam deflection range: > +/- 5 degrees

Quad Detectors

Quad detectors are commonly used in laser beam tracking applications. Most quad detectors are silicon based, and they respond to light in the visible and near IR-spectrum. However, quad detectors can be made from different material systems; therefore, they cover various spectral ranges.

The incoming light from the remote location is focused onto the detectors by using an external optic, such as a lens or a mirror. The detector consists of four separate single-detector elements that arrange in a matrix. Each of these four elements collects the light separately. If the spot is located exactly in the middle of the detector array, the signal output from all four detectors is the same. If the light spot moves, the amount of light collected by each different detector will be different, resulting in a different level of the output signal. By analyzing and comparing these four individual output signals, you can determine the direction of spot movement on the surface of the detector array. Because only four different output levels have to be analyzed, this procedure can be fast.

Among other factors, the pointing resolution of this method depends on the size and spacing of the detector elements in the imaging plane. If the detector elements are small, the overall resolution is higher. With appropriate amplifier combinations, light spot movements of 10 µm or even less can be detected. However, the total vertical and horizontal field of regard is determined by the size of the detector matrix. If the light spot leaves the detector surface, the system loses its tracking capability. Therefore, quad detectors are normally used to track small angle deviations (fine tracking).

CCD Arrays

Charge-coupled devices (CCD) can be found in many commercial applications that require the conversion of an optical signal into an electronic signal. CCDs are the heart of modern cameras and camcorders. Two-dimensional arrays with large pixel count numbers (256×from 256 up to 2048×2048) are commercially available. CCD cameras—such as Web cameras—with a 512×512 array chip and a completely integrated and packaged readout system including optics can be found in stores for less than $100.

CCD pixel sizes can be small (10 μm), with an even much smaller gap between the detector elements. Because of the greater number of pixels, the total detection area is large when compared to a quad detector. The larger detection area automatically translates into a wider field of view for tracking applications. However, because the readout system for CCD chips is based on a serial shift register approach, you must read the information for the complete chip, even if the actual region of interest (ROI), the spot location, covers only a small fraction of the total chip.

Commercial tracking systems that use a CCD position detector often incorporate a computer-controlled frame grabber card. The software running on the computer performs the actual position detection and feeds this information back to the system for counter measures such as moving a steering mirror. This process can take quite some time, so the tracking can be too slow to follow the movement of the light beam.

A better approach includes the use of a digital signal processor (DSP) to perform the position analysis and the feedback control. However, even though DSPs can perform an image analysis effectively, this approach is also limited by the readout time of the CCD chip. Larger pixel count, direct-readout detector arrays offering the capability to read individual pixels separately, would solve this problem. However, most direct-readout CCD chips are still in the development stage.

Link Margin Analysis

A detailed link margin analysis is an important engineering task for any communication system. For example, for optical fiber-based systems, the engineer looks at the amount of power that is launched into the fiber at the transmitter side and then determines all potential losses until the signal arrives at the receiver side. In fiber systems, the fiber, connectors, slicers, and so on cause these losses. The receiver typically has a specific minimum sensitivity at a given data rate, and the task is to make sure that the launch power minus the loss factor stays above the minimum sensitivity to guarantee reliable operation of the system. This procedure is similar for FSO systems.

However, an important difference between fiber-based and FSO communication systems is related to the fact that the loss of the media (air) between the transmitter and the receiver can vary in time due to the impact of weather. Therefore, it is important for FSO systems to take weather conditions into consideration. The following sections explain the link margin analysis for FSO systems in more detail.

Imagine a communications link that goes from a transmitter at point A to a receiver at point B, and a signal that is being sent between them. Figure 2.13 shows the picture of a typical FSO installation on a high-rise in a metropolitan downtown area. How far apart can the transmitter and receiver be to still be able to communicate?

FIGURE 2.13
Typical setup of an FSO communication link on a high-rise building in a metropolitan downtown area.

To answer this question, we need to look at the signal power available at the receiver after the signal reaches the opposite location. It is also important to understand how well the receiver utilizes this power, and more importantly, how reliable the communication link will be over the anticipated distance. This last factor is critically important for free-space optical links because the received signal power can vary significantly over time.

To quantify these factors, we have to build a *link budget* for our communications system, totaling the expected gains and losses, and then comparing the received signal power to the level required by the signal-detecting circuitry. Any excess of power is dubbed the *link margin*; it is a measure of how much leeway, or buffer, we have in the event that undesirable time-varying channel effects (*fading*) cause a deterioration in the power at the receive side.

To make the link more reliable, we would want to leave a larger link margin than the minimum nominal received power level so that a larger buffer (*fade margin*) is available. However, allocating signal power to the link margin means taking away power that would otherwise go to increasing the distance at which we could separate the transmitter and receiver. This simple logic shows a trade-off between distance and reliability.

Consider a simple example of a link budget: Assume that we have a power output from the transmitter of 4 mW, which corresponds to 6 decibels per milliwatt (dBm) on the frequently used decibel scale. This value will be our starting point for the link budget calculation. To figure out how much of this power arrives at the receiver, we need to look at the losses in the communications channel.

Optical Loss

The first source of loss in a free-space optical system is due to imperfect lenses and other optical elements (such as couplers). For example, a lens might transmit 96% of the light, but 4% gets reflected or absorbed and is no longer available. To take this into account, we will put a line called "optical loss" in the link budget calculation. The amount of loss depends on the characteristics of the equipment and quality of the lenses. This value needs to be measured or derived from the manufacturer of the optical components. In our example, the optical loss has been measured as a 4 dB reduction in signal power. Consequently, we will subtract 4 dB from the original 6 dBm that we started with at the beginning:

$$6 \text{ dBm} - 4 \text{ dB} = 2 \text{ dBm}$$

To develop the link budget, it is helpful to create a spreadsheet that lists all the quantities we have to add or subtract in the link budget calculation. Such a spreadsheet is shown in Table 2.4.

TABLE 2.4 The First Part of the Link Budget

Description	Value	Units
Transmit Power	6	dBm
Optical Losses	−4	dB

However, in a typical free-space optical system, several other sources of loss can be found in the communication channel. These include geometrical loss, pointing loss, and atmospheric loss.

Geometrical Loss

The term *geometrical loss* refers to the losses that occur due to the divergence of the optical beam. Typically, a free-space optical system is engineered such that the beam diverges by some amount over the path from the transmitter to the receiver. In some systems that use active tracking, this divergence can be quite small. In systems that do not use active tracking or in which the tracking system is in the several-hertz range, the divergence of the beam is engineered so that when the beam wobbles, some part of it will always hit the receiver, and the link will be maintained.

The result of divergence is that much, or even most, of the light is never collected by the receiver. The loss is equal to the area of the receiver collecting optics relative to the area of the beam at the receiver. For a single beam, the area of the beam at the receiver can be calculated using a simple geometrical formula, assuming that the divergence occurs at a constant rate as soon as the beam leaves the transmitter. Figure 2.14 shows the projected beam diameter for a 4 mrad beam at distances of 300, 1,000, and 2,000 meters. At these projected distances, the beam continuously increases from 1.3 meters to 4.0 meters and, finally, 8 meters, respectively.

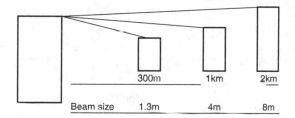

FIGURE 2.14
Projected beam sizes for a 2 mrad divergence angle.

The assumption of a linear beam spread is pretty accurate because most systems are designed to operate under conditions where Rayleigh propagation does not have to be taken into account. Therefore, the ratio of the projected beam size areas and the receive optics area is simply this:

> Ratio = [Diameter of Receive Optics / (Diameter of Transmit Optics + Distance * Divergence angle)]^2

If we measure the diameters in cm, the distance in km, and the divergence in mrad, the formula becomes the following:

$$\frac{A_R}{A_B} = \left[\frac{D_R}{D_T + 100 \ ^* d \ ^* \theta} \right]^2$$

in which A_R is the Area of the Receiver, and A_B is the Area of the Beam. We can express this quantity in dB also, so that it is compatible with the first part of our link budget. As an example, if the beam diameter at the transmitter is 3 cm, the receiver lens is 8 cm, and the divergence is 2 mrad, the loss at 1 km can be calculated by using the previous equation. The result of this calculation yields the following:

$$A_R/A_B = 0.0015685 = -28 \cdot dB$$

For systems that use multiple, overlapping beams to transmit the data or for beam profiles that are not uniform, the calculation of geometrical loss becomes more complicated. However, the basic principle remains the same, and an analytic geometrical formula can be developed. The answer regarding the geometrical path loss can also be found by using numerical integration. A multibeam approach has proven to be successful in fighting an atmospheric effect called scintillation that will be discussed in more detail in Chapter 3.

The geometrical loss that we calculate is added as another line item to the link budget, which is shown in Table 2.5. To calculate the geometrical path loss, we will choose a distance of 1 km.

TABLE 2.5   The Link Budget with the Geometrical Loss Added

Description	Value	Units
Transmit Power	6	dBm
Optical Losses	–4	dB
Geometrical Loss	–28	dB

Pointing Loss

Perhaps the distance between the transmitter and receiver is large enough that during installation of the link, it is difficult to see the far side. Or perhaps the tracking system contains residual steady-state errors. If either of these conditions exists, additional loss can be incurred because the transmitter is not pointed accurately enough at the receiver. Typically, these effects are seen for distances in excess of 3 km, at which point we might subtract an additional dB of power from the link budget. In our example, we have set the link distance at 1 km. Therefore, we will not consider the pointing loss to be significant, and we will set it at 0 dB. However, for all practical purposes, the pointing loss will add another line item to our link budget table. The updated table is shown as Table 2.6.

TABLE 2.6 The Link Budget with the Pointing Loss Added

Description	Value	Units
Transmit Power	6	dBm
Optical Losses	−4	dB
Geometrical Loss	−28	dB
Pointing Loss	0	dB

Atmospheric Loss and Receiver Sensitivity

The atmosphere causes signal degradation and attenuation in a free-space optical system link in several ways, including absorption, scattering (mainly Mie scattering), and scintillation. All these effects are time varying and will depend on the current local conditions and weather. All of these elements contribute to *channel fade*; they will be explained in more detail in Chapter 3.

The final goal of the link budget calculation is to examine how far we can place the transmitter and receiver while still maintaining enough margin to allow for a specified minimum link availability (for example, 99.9%). If we choose a distance, then we want to know what the link fade margin is at this distance. From this value, we can judge the reliability of the link. In our example, we will find the link margin, and this will give us a quantitative value for the amount of atmospheric loss the system will be able to tolerate.

The receiver sensitivity is a measure of how well the signal detection circuitry can make use of the received power level. In FSO systems, simple binary encoding and on-off keying is used. This means that a "1" bit is light on, and a "0" bit is light off. The receiver must be able to detect these two different binary states. For different receiver types, a theoretical limit exists for how low the signal power can be and still be visible above the background noise.

The receiver sensitivity is also a function of modulation speed of the incoming signal: Higher speed signals (shorter bits in time) contain less photons that can be detected by the receiver, making it more difficult to resolve a "1" or a "0" state. Different receiver designs approach this limit with varying degrees of success.

The background noise can be from several sources, such as ambient light, *shot noise* (noise caused by random fluctuations in the motion of charge carriers in a conductor), and *thermal noise* (noise generated by thermal agitation of electrons in a conductor). With an Avalanche Photo Diode (APD), the noise will be increased during the amplification process (*multiplicative noise*). From an equipment design perspective, it is important to know the strengths of the various noise sources to calculate what the expected sensitivity should be, and finally compare this with the fabricated equipment. However, from a systems design point of view, the equipment supplier will provide the sensitivity figure of a particular receiver. If the data is not available, it can be measured using an optical power meter and a Bit Error Rate (BER) tester.

The specified receiver sensitivity will apply to a specific Bit Error Rate. The effect of noise causes increases in the Bit Error Rate until it exceeds some specific and predefined threshold. The threshold that is chosen will depend on the specific application, and for high-rate data transmission, a BER threshold of better than 1×10^{-10} is often used. For our example, we will assume that the equipment supplier has specified the receiver sensitivity for our equipment at −43 dBm for a 155 Mbps digital transmission rate and a BER of 1×10^{-10}. This number is entered as the last line in our link budget, and subtracted from the others so that we can then calculate the link margin. The completed version of the link budget spreadsheet is shown in Table 2.7.

TABLE 2.7 The Completed Link Budget Showing the Link Margin

Description	Value	Unit
Transmit Power	6	dBm
Optical Losses	−4	dB
Geometrical Loss	−28	dB
Pointing Loss	0	dB
Link Margin	17	dB

The 17 dB of link margin that we calculated for this specific application is what is available to use to protect against fading events caused by the atmosphere.

Simple Link Analysis Tool

To allow you to perform some basic link margin calculations, LightPointe generated a generic spreadsheet that can be downloaded free from their Web site (http://www.lightpointe.com).

To run the spreadsheet, the user will need Microsoft Excel.

Summary

Free-space optics system components basically consist of a light source, optics to direct and focus it, receive capabilities, and electro-optic electronics to handle conversions of electronic and optical communications. The technologies required are, for the most part, similar to conventional fiber optics with the exception of some unique requirements caused by using "free space" rather than fiber strands as the transmission medium. Those unique requirements are more thoroughly discussed in Chapter 3, "Factors Affecting FSO."

Factors Affecting FSO

IN THIS CHAPTER

At one time, connecting all of the people all of the time around all of the world was a nice idea but completely impractical. The Internet has changed all of that, and the possibility now exists.

How about all the bandwidth desired for all the high-bandwidth users in all the land? Can free-space optics deliver on *this* proposition? Well, if it weren't for fog (and other assorted atmospheric and installation-related issues) the light beams of FSO might just be *that* "silver bullet." As it is, FSO, although a bullet indeed, is perhaps a brass-jacketed one.

As with most technologies, knowledge is power. And armed with the knowledge of FSO's enemies, you will possess the power to properly deploy FSO where it is the right choice. You will also be capable of avoiding the chasm of "right tool, wrong application," and thus avoid incorrect selection when it is nonoptimal. This chapter discusses the factors that can affect the viability and success of FSO.

Transmission of IR Signals Through the Atmosphere

Even a clean, clear atmosphere is composed of oxygen and nitrogen molecules. The weather can contribute large amounts of water vapor. Other constituents can exist, as well, especially in polluted regions. These particles can scatter or absorb infrared photons propagating in the atmosphere.

Although it is not possible to change the physics of the atmosphere, it is possible to take advantage of optimal atmospheric windows by choosing the transmission wavelengths accordingly. To ensure a minimum amount of signal attenuation from scattering and absorption, FSO systems operate in atmospheric windows in the IR spectral range. As discussed in Chapter 2, "Fundamentals of FSO Technology," today's commercially available FSO systems operate in the near IR spectral windows located around 850 nm and 1550 nm. Other windows exist in the wavelength ranges between 3–5 μm and 8–14 μm. However, their commercial use is limited by the availability of devices and components and difficulties related to the practical implementation such as low-temperature cooling.

The impact of scattering and absorption on the transmission of light through the atmosphere is discussed in more detail in the following sections.

Beer's Law

Beer's Law describes the attenuation of light traveling through the atmosphere due to both absorption and scattering. In general, the transmission, τ, of radiation in the atmosphere as a function of distance, x, is given by Beer's Law, as

$$I_R/I_0 = \tau = \exp(-\gamma\,x)$$

where I_R/I_0 is the ratio between the detected intensity I_R at the location x and the initially launched intensity I_0, and γ is the attenuation coefficient.

The attenuation coefficient is a sum of four individual parameters—molecular and aerosol scattering coefficients α and molecular and aerosol absorption coefficients β—each of which is a function of the wavelength. You will see the application of this relationship among received intensity, scattering, and absorption a little later in this chapter.

The attenuation coefficient is given as

$$\gamma = \alpha_m + \alpha_a + \beta_m + \beta_a$$

This formula shows that the total attenuation, represented by the attenuation coefficient γ, results from the superposition of various scattering and absorption processes. This will be discussed in more detail in the following sections.

Scattering

Scattering refers to the "pinball machine" nature of light trying to pass through the atmosphere. Light scattering can drastically impact the performance of FSO systems. Scattering is not related to a loss of energy due to a light absorption process. Rather, it can be understood as a redirection or redistribution of light that can lead to a significant reduction of received light intensity at the receiver location. A nice overview of these processes can be found in the literature[1].

Several scattering regimes exist, depending on the characteristic size of the particles, (r), the light encounters on the trip to its destination. One description is given as $x_0 = 2\pi r/\lambda$, where λ is the transmission wavelength and r is particle radius. For $x_0 \ll 1$, the scattering is in the Rayleigh regime; for $x_0 \approx 1$, the scattering is in the Mie regime; and for $x_0 \gg 1$, the scattering can be handled using geometric optics. Compared to infrared wavelengths usually used in free-space optics, the average radius of fog particles is about the same size. This is the reason that fog is the primary enemy of the beam. Rain and snow particles, on the other hand, are larger, and thus present significantly less of an obstacle to the beam.

Rayleigh Scattering

A radiation incident on the bound electrons of an atom or molecule induces a charge imbalance or dipole that oscillates at the frequency of the incident radiation. The oscillating electrons reradiate the light in the form of a scattered wave. Rayleigh's classical formula for the scattering cross section is as follows:

$$\sigma_s = \frac{f e^4 \lambda_0^4}{6\pi \varepsilon_0^2 m^2 c^4} \frac{1}{\lambda^4}$$

where f is the oscillator strength, e is the charge on an electron, λ_0 is the wavelength corresponding to the natural frequency, $\omega_0 = 2\pi c/\lambda_0$, ε_0 is the dielectric constant, c is the speed of light, and m is the mass of the oscillating entity. The λ^{-4} dependence and the size of particles found in the atmosphere imply that shorter wavelengths are scattered much more than longer wavelengths. Rayleigh scattering is the reason why the sky appears blue under sunny weather conditions. However, for FSO systems operating in the longer wavelength near infrared wavelength range, the impact of Rayleigh scattering on the transmission signal can be neglected. The wavelength dependence of the Rayleigh scattering cross section in the infrared spectral range is shown in Figure 3.1.

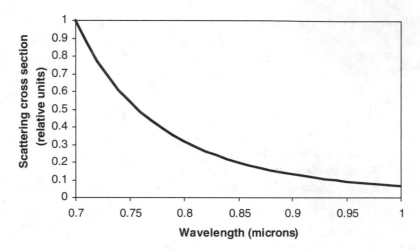

FIGURE 3.1
Rayleigh scattering cross section versus infrared wavelength.

Mie Scattering

The Mie scattering regime occurs for particles about the size of the wavelength. Therefore, in the near infrared wavelength range, fog, haze, and pollution (aerosols) particles are the major contributors to the Mie scattering process. The theory is complicated, but well understood. The problem arises in comparing the theory to an experiment. Because absorption dominates most of the spectrum, data must be collected in wavelength ranges that occur in an atmospheric window, with the assumption that only scattering is taking place. In addition, the particle distributions must be known. For aerosols, this distribution depends on location, time, relative humidity, wind velocity, and so on. An empirical simplified formula that can be found in literature [1] and that is used in the FSO community for a long time to calculate the attenuation coefficient due to the Mie scattering is given by the following:

$$\gamma = \frac{3.91}{V} \left(\frac{\lambda}{550}\right)^{-\delta}, \text{ where } \delta = 0.585(V)^{1/3} \text{ for } V < 6\text{Km}$$

$$\delta = 1.6 \qquad \text{for } V > 50 \text{ Km}$$
$$\delta = 1.3 \qquad \text{for } 6 \text{ Km} < 50 \text{ Km}$$

In this formula, V corresponds to the visibility, and λ is the transmission wavelength. However, this formula has been challenged recently by the FSO research community. The transmission wavelength dependency of the attenuation coefficient γ does not follow the predicted empirical formula. More precise numerical simulations of the exact Mie scattering formula suggest that the attenuation coefficient does not drastically depend on wavelength as far as the near infrared wavelength range typically used in FSO systems is concerned. The overall conclusion that can be derived from empirical observation is that Mie scattering caused by fog characterizes the primary source of beam attenuation, and that this effect is geometrically accentuated as distance is increased. For all practical purposes, the visibility conditions in the FSO deployment area must be studied. Visibility data collected over several decades is available from the National Weather Services and can be used to derive distance-dependent availability figures for a particular geographic region of deployment. However, a complication results from the fact that weather conditions are typically measured at airports that can be located away from the actual FSO installation location. Some FSO vendors have started to collect data directly from metropolitan areas and cross-correlate these findings with data collected at nearby airports to optimize the availability statistics. Environments with strong variations in microclimate are especially challenging. For most commercial FSO deployments, operation in heavy fog environments requires keeping the distances between FSO terminals short to maintain high levels of availability. The link power margins of most vendor equipment allow for availabilities that exceed 99.99% if distances are kept below 200 m.

Absorption

Atoms and molecules are characterized by their index of refraction. The imaginary part of the index of refraction, k, is related to the absorption coefficient, α, by the following:

$$\alpha = \frac{4\pi k}{\lambda} = \sigma_a N_a$$

where σ_a is the absorption cross section and N_a is the concentration of the absorbing particles. In other words, the absorption coefficient is a function of the absorption strength of a given species of particle, as well as a function of the particle density.

Atmospheric Windows

In the atmospheric window most commonly used for FSO, infrared range, the most common absorbing particles are water, carbon dioxide, and ozone. A typical absorption spectrum is shown in Figure 3.2. Vibrational and rotational energy states of these particles are capable of absorption in many bands. Well-known windows exist between 0.72 and 15.0 μm, some with narrow boundaries. The region from 0.7–2.0 μm is dominated by water vapor absorption, whereas the region from 2.0–4.0 μm is dominated by a combination of water and carbon dioxide absorption.

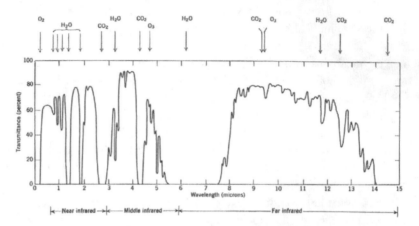

FIGURE 3.2
Atmospheric transmittance measured over a sea level 1820 m horizontal path [2].

Atmospheric Absorbers

The abundance of absorbing species determines how strongly the signal will be attenuated. These species can be broken up into two general classes: molecular and aerosol absorbers. Figure 3.3 shows the transmission spectrum for clear sky conditions with a standard urban aerosol concentration providing a visibility of 5.0 km. This graph was generated by using the Air Force's MODTRAN[3] program. Included in this calculation was absorption from water vapor, carbon dioxide, and so on.

In the near infrared, water vapor is the primary molecular absorber, with many absorption lines to attenuate the signal. Above 2.0 μm, both water vapor and carbon dioxide play a large role. The vibrational and rotational transitions determine which energies are easily absorbed, but the large number of permutations greatly increases the number of lines. Figure 3.4 shows the clear sky transmission for water vapor only. You can see that water vapor dominates the clear sky transmission in the near infrared. The large number of lines contributes to a complicated spectrum with occasional windows at popular FSO frequencies, such as 850 and 1,550 nm. Figure 3.4 shows the carbon dioxide transmission. Occasional sharp resonant peaks are superimposed on an overall relatively flat background.

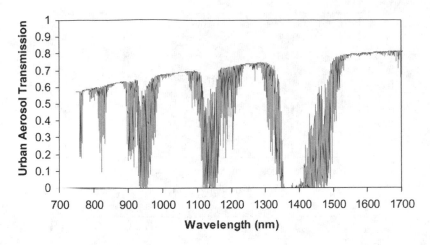

FIGURE 3.3

Transmission as a function of wavelength under urban aerosol conditions (visibility = 5 km), as calculated by MODTRAN.

Aerosols occur naturally in the form of meteorite dust, sea-salt particles, desert dust, and volcanic debris. They can also be created as a result of man-made chemical conversion of trace gases to solid and liquid particles and as industrial waste. These particles can range in size from fine dust less than 0.1 μm to giant particles greater than 10.0 μm. One estimate determined that 80% of the aerosol mass is contained within the lowest kilometer of the atmosphere. Land produces more aerosols than ocean, and the Northern Hemisphere produces 61% of the total amount of aerosols in the world.[4] Because the radii span the infrared, scattering from these particles can definitely be a problem for FSO systems. However, these particles also absorb in the infrared wavelengths. For example, carbon and iron have many absorption lines, but their abundance in the atmosphere is usually limited. Figure 3.5 shows the clear sky transmission including urban aerosols. A comparison of Figures 3.5 and 3.4 shows how the transmission of the atmosphere is affected by aerosol particles.

Turbulence

The desert might seem the perfect location for an FSO system. This is certainly true as far as the attenuation of the atmosphere is concerned. However, in hot, dry climates, turbulence might cause problems with the transmission. As the ground heats up in the sun, the air heats up as well. Some air cells or air pockets heat up more than others. This causes changes in the index of refraction, which in turn changes the path that the light takes while it propagates through the air. Because these air pockets are not stable in time or in space, the change of index of refraction appears to follow a random motion. To the outside observer, this appears as turbulent behavior.

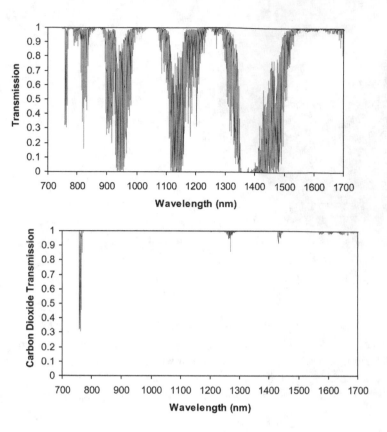

FIGURE 3.4

Clear sky transmission as a function of wavelength for water (top) and carbon dioxide (bottom) as calculated by MODTRAN.

Laser beams experience three effects under turbulence. First, the beam can be deflected randomly through the changing refractive index cells. This is a phenomenon known as *beam wander*. Because refraction through a media such as air works similarly to light passing through any other kind of refractive media such as a glass lens, the light will be focused or defocused randomly, following the index changes of the transmission path. Second, the phase front of the beam can vary, producing intensity fluctuations or scintillation (heat shimmer). Third, the beam can spread more than diffraction theory predicts.[1]

A good measure of turbulence is the refractive index structure coefficient, Cn^2. Because the air needs time to heat up, the turbulence is typically greatest in the middle of the afternoon ($Cn^2 = 10^{-13}$ m$^{-2/3}$) and weakest an hour after sunrise or sunset ($Cn^2 = 10^{-17}$ m$^{-2/3}$). Cn^2 is usually largest near the ground, decreasing with altitude. To minimize the effects of scintillation on the transmission path, FSO systems should not be installed close to hot surfaces. Tar roofs, which can experience a high amount of scintillation on hot summer days, are not preferred installation spots. Because scintillation decreases with altitude, it is recommended that FSO systems be installed a little bit higher above the rooftop (>4 feet) and away from a side wall if the installation takes place in a desert-like environment.

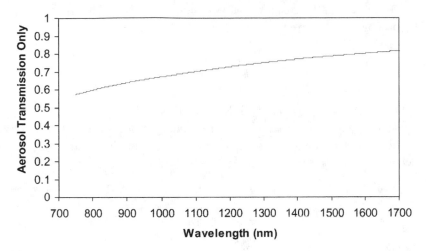

FIGURE 3.5

Transmission as a function of wavelength for urban aerosol only as calculated by MODTRAN.

Beam Wander

For a beam in the presence of large cells of turbulence compared to the beam diameter, geometrical optics can be used to describe the radial variance, σ_r, as a function of wavelength and distance, L, as follows:

$$\sigma_r = 1.83 C_n^2 \lambda^{-1/6} L^{17/6}$$

This relationship implies that longer wavelengths will have less beam wander than shorter wavelengths, although the wavelength dependence is weak. Although keeping a narrow beam on track might be a problem, the rate of fluctuations is slow (under a kHz or two), such that a tracking system can be used.

Scintillation

When you have seen a mirage that appears as a lake in the middle of a hot asphalt parking lot, you have experienced the effects of atmospheric scintillation. Of the three turbulence effects, free-space optical systems might be most affected by scintillation. Random interference with the wave front can cause peaks and dips, resulting in receiver saturation or signal loss. "Hot spots" in the beam cross section can occur of the size $\sqrt{\lambda L}$, about 3 cm for an 850 nm beam

1 Km away. A great deal of work was done on this topic for applications like telescope signals and earth-satellite links, where a majority of the scintillation could be observed near the Earth's surface. FSO systems operate horizontally in the atmosphere near the surface, experiencing the maximum scintillation possible.

Scintillation effects for small fluctuations follow a log-normal distribution, characterized by the variance, σ_i, for a plane wave given by the following:

$$\sigma_i^2 = 1.23 C_n^2 k^{7/6} L^{11/6}$$

where $k = 2\pi/\lambda$. This expression suggests that larger wavelengths would experience a smaller variance, all other factors being equal. For FSO systems with a narrow, slightly diverging beam, the plane wave expression is more appropriate than that for a spherical beam. Even if the wave front is curved when it reaches the detector, the transmitting beam is so much larger than the detector that the wave front would be effectively flat.

The expression for the variance for large fluctuations is as follows:[5]

$$\sigma_{high}^2 = 1.0 + 0.86 \left(\sigma^2 \right)^{-2/5}$$

suggesting that shorter wavelengths would experience a smaller variance. In FSO deployment, the beam path must be more than 5 m above city streets or other potential sources of severe scintillation.

Beam Spreading

The beam size can be characterized by the effective radius, a_t, the distance from the center of the beam ($z = 0$) to where the relative mean intensity has decreased by $1/e$. The effective radius is given by the following:

$$a_t = 2.01 \left(\lambda^{-1/5} C_n^{6/5} z^{8/5} \right)$$

The wavelength dependency on beam spreading is not strong. The spot size can often be observed to be twice that of the diffraction-limited beam diameter. Many FSO systems incur approximately 1 m of beam spread per kilometer of distance. In a perfect world with no environmental attenuators present, beam spread would be the only distance-limiting variable.

The Impact of Weather

So far, the discussions in this chapter have been somewhat theoretical. One of the practical topics of most interest to designers and implementers of FSO systems is the weather.

Rain

Rain has a distance-reducing impact on FSO, although its impact is significantly less than that of fog. This is because the radius of raindrops (200–2000 μm) is significantly larger than the wavelength of typical FSO light sources.

Typical rain attenuation values are moderate in nature. For example, for a rainfall of 2.5 cm/hour, a signal attenuation of 6 dB/km can be observed. Therefore, commercially available FSO systems that operate with a 25 dB link margin can penetrate rain relatively unhindered. This is especially the case when systems are deployed in metropolitan areas where building distances are typically much less than 1 km. If, for example, the system is deployed over a distance of 500 m under the same rain conditions, the attenuation is only 3 dB/km. However, when the rain rate increases dramatically to beyond the cloudburst level (> 10 cm/hour), rain attenuation can become an issue in deployments beyond the distance scale of a typical metropolitan area. However, these kind of cloudbursts last for only a short period of time (minutes).

An interesting point to note is that RF wireless technologies that use frequencies above approximately 10 GHz are adversely impacted by rain and little impacted by fog. This is because of the closer match of RF wavelengths to the radius of raindrops, both being larger than the moisture droplets in fog. The lower unlicensed RF frequencies in the 2.4 GHz and 5.8 GHz ranges are relatively unaffected by rain or fog, but incur significant interference risks by nature of the lack of licensing in those frequencies.

Snow

Snowflakes are ice crystals that come in a variety of shapes and sizes. In general, however, snow tends to be larger than rain. Whiteout conditions might attenuate the beam, but scattering doesn't tend to be a big problem for FSO systems because the size of snowflakes is large when compared to the operating wavelength. The impact of light snow to blizzard and whiteout conditions falls approximately between light rain to moderate fog, with link attenuation potentials of approximately 3 dB/km to 30 dB/km.

Fog

Fog is the most detrimental weather phenomenon to FSO because it is composed of small water droplets with radii about the size of near infrared wavelengths. The particle size distribution varies for different degrees of fog. Weather conditions are typically referred to as fog when visibilities range between 0–2,000 meters. Because foggy conditions are somewhat difficult to describe by physical means, descriptive words such as "dense fog" or "thin fog" are sometimes used to characterize the appearance of fog. When the visibility is more than 2,000 meters, the condition is often referred to as hazy.

Table 3.1 relates visibility and different fog conditions. Scattering is the dominant loss mechanism for fog. Even modest fog conditions can highly attenuate infrared signals over shorter distances. The expected path attenuation in dB/km and its correlation to visibility is shown in the table. The table also clearly illustrates that rain has much less impact on FSO systems' path losses when compared to fog. For example, a medium rainfall results in less attenuation than a thin fog.

TABLE 3.1 International Visibility Codes for Weather Conditions and Precipitation

Weather Condition	Precipitation	Amount mm/hr	Visibility	dB Loss/km
Dense fog			0 m, 50 m	−271.65
Thick fog			200 m	−59.57
Moderate fog	snow		500 m	−20.99
Light Fog	snow	Cloudburst	100	770 m	−12.65
				1 km	−9.26
Thin fog	snow	Heavy rain	25	1.9 km	−4.22
				2 km	−3.96
Haze	snow	Medium rain	12.5	2.8 km	−2.58
				4 km	−1.62
Light haze	snow	Light rain	2.5	5.9 km	−0.96
				10 km	−0.44
Clear	snow	Drizzle	0.25	18.1 km	−0.24
				20 km	−0.22
Very Clear			23 km	−0.19
				50 km	−0.06

Fog is not well understood, and it is difficult to characterize physically. Although visibility is most commonly used to characterize foggy conditions, other methods such as particle size and density measurements have been undertaken to describe fog conditions in a more quantitative way. The FSO community mainly uses visibility data because these measurements have been taken at major airports over many decades. To some extent, these measurements allow you to characterize different regions and derive statistical availability figures for FSO systems. However, most of the data has been time averaged over years; in general, the temporal resolution of these data points is not very high.

Because microclimate environments such as ponds or rivers can induce foggy conditions, the data taken at airports sometimes is not reliable for nearby environments. However, it has been shown that the visibility at airports provides a good estimate for the minimum expectable availability figure. This is because airports typically are located outside metropolitan boundaries, and the microclimate inside a city typically generates less foggy conditions.

The density distribution of fog particles can also vary with height, which makes the modeling of fog even more complex. The limited amount of information regarding the local impact of fog on the availability of FSO systems is certainly one of the biggest challenges for the FSO industry.

Line of Sight (LOS)

FSO system operation requires line of sight (LOS). Line of sight simply means that the transmitter and the receiver at both networking locations can see each other. Because IR beams propagate and expand in a linear fashion, the line of sight criteria is less strict when compared to microwave systems that require an additional path clearance to account for the extension of Fresnel zones.

Determining LOS

The easiest way to find out if line of sight exists between two remote locations is visual observation. For distances that are longer than a mile, this might not be trivial. Field glasses and telescopic lenses might be necessary in these scenarios. Many FSO vendors incorporate an alignment telescope into the FSO terminals to accomplish this task. Some organizations prefer to use more sophisticated GIS maps before sending a field crew out to assess a site. A variety of GIS mapping software vendors exist whose programs can load high-resolution and three-dimensional topology maps. These maps include information regarding buildings and their specific locations. This allows for determining whether line of sight exists between two known locations.

Although rooftop to rooftop is one of the more typical deployment scenarios for FSO, it might be possible to locate the transceivers behind windows in the building when roof access is not available. However, you must take care to determine whether line of sight can be achieved. In addition, the angle the beam makes with the window is critical. The angle should be as perpendicular as possible, yet slightly angled (5 degrees) to reduce bounce-back of the beam to its own receptor. At some angle, no light will be transmitted at all. This complete internal reflection is what keeps light inside fiber-optic cable. Also, some windows contain glass or glass coatings that reduce glare. Because these windows are often specifically designed to reject infrared, the coatings can reduce the signal by 60% or even more. Sometimes, window connections have no alternatives because roof rights cannot be obtained. Decreasing the link distance (which increases the power of the signal at the receiving telescope due to decreased geometrical loss) or increasing beam intensity can often solve the problem.

Other Factors Affecting FSO

When planning an FSO deployment, you must consider the application intended. Is the data traffic low-speed overnight downloads or high-speed uninterruptible video data? Is the distance between sites long? Is the location notoriously foggy? These factors all influence the selection of the most appropriate FSO system.

Visibility

Low visibilities will decrease the effectiveness and availability of FSO systems. Long-term weather observations show that some cities, such as Seattle, WA, have lower average visibilities than cities such as Denver, CO. This means that for the same distance, the same FSO system in Denver will experience a higher availability than a system installed in Seattle. Low visibility can occur during a specific time period within a year or at specific times of the day (such as in the early morning hours). Especially in coastal areas, low visibility can be a localized phenomena (coastal fog). This means that for the same distance, the same FSO system in Denver will experience less downtime than in Seattle.

One solution to the negative impact of low visibility is to shorten the distance between FSO terminals to maintain a specific statistical availability figure. This provides a greater link margin to handle bad weather conditions such as dense fog. Redundant path operation can improve the availability if the visibility is limited on a local scale. Examples are fog across a river or pond or an air conditioner's exhaust stream on top of a roof. Another solution is to use a multiple beam system to maintain a higher link availability.

Low visibility and the associated high scattering coefficients are the most limiting factors for deploying FSO systems over longer distances.

Distance

Distance impacts the performance of FSO systems in three ways. First, even in clear weather conditions, the beam diverges and the detector element receives less power. For a circular beam, the geometrical path loss increases by 6 dB when the distance is increased by a factor of two. Second, the total transmission loss of the beam increases with increasing distance. Third, scintillation effects accumulate with longer distances. Therefore, the value for the scintillation fade margin in the overall power budget will increase to maintain a predefined value for the BER.

Most commercially available FSO systems are rated for operation between 25–5,000 m, with high-powered military and satellite systems capable of up to 2,000 km. Most systems rated for greater than 1 km incorporate three or more lasers operating in parallel to mitigate distance-related issues. It is interesting to note that in the vacuum of space, FSO can achieve distances of thousands of kilometers.

Bandwidth

In standard O-E-O FSO systems, two elements limit the bandwidth of the overall system. These elements are the transmission source and the photo detector. When LEDs are incorporated into FSO systems, the bandwidth is typically limited to 155 Mbps. When laser sources are used, the speed can be much higher. Directly modulated lasers operating up to 2.5 Gbps are commercially available for use in FSO systems. At higher speed such as 10 Gbps or above, external modulators can be used to modulate the cw output of a laser source.

With respect to the photo detector, inexpensive Si-Pin diodes and Si-APDs supporting data rates up to 1,250 Gbps are commercially available. For operation in the 1.5 micrometer wavelength band, InGaAs detectors are used. Commercially available and off-the-shelf detectors that support a bandwidth of 10 Gbps and beyond can be used in FSO systems. However, at higher bit rates (shorter bit durations), the amount of light that can be collected by the receiver and converted into electrons is extremely low and the sensitivity of a receiver becomes a function of the bit rate. In general terms, this means the higher the bit rate, the less sensitivity. Typical sensitivity ratings are –43dBm@155Mbps and –34dBm@622Mbps. When the system reaches its sensitivity limit, the thermal (Johnson) noise impacts the bit error rate (BER) of the system.

Selecting the Transmission Wavelength

To select the best wavelength to use for free-space optical communication systems, you must consider several factors. In general, the specific wavelength is not so important as long as the transmission wavelength does not correspond to a wavelength that is strongly absorbed in the atmosphere. As stated previously, Mie scattering is by far the dominant factor as far as attenuation of an IR signal through the atmosphere is concerned. However, applications in dense urban areas with high aerosol contents might slightly benefit from a different wavelength than relatively unpolluted suburban locations.

As has been mentioned, some atmospheric advantages exist for some wavelengths being used for FSO systems, but that is not the whole picture. Another issue has to do with the fact that at approximately 1,550 nm, the regulatory agencies allow approximately 100 times higher power for "eye safe" lasers. This is because at this wavelength, the aqueous fluid of the eye absorbs much more of the energy of the beam, preventing it from traveling to the retina and inflicting damage. The disadvantage of this laser type is mainly cost when compared to shorter wavelength lasers operating around 850 nm. Design engineers must deal with the cost of implementing such a system.

Choosing the correct transmission wavelength involves many factors, such as availability of components, price, required transmission distance, eye-safety considerations, and so on. As noted at the beginning of this chapter, the preferred wavelengths are in the 850 nm and 1550 nm wavelength band. Operation in the longer wavelength transmission windows between 3–5 μm and 8–14 μm has also been suggested by the FSO community due to the excellent transmission characteristics of the atmosphere in the mid infrared wavelength range. However, some recent more detailed studies of the Mie scattering coefficients in the mid infrared range suggest that there is no significant advantage in using longer IR-wavelength such as 3.5 μm instead of the 850 or 1,550 nm wavelength ranges to counteract scattering losses. Also, the availability of components such as light sources and detectors in the mid IR wavelength range is very limited. At present, most highly sensitive detectors and light sources in this wavelength range must be cooled to low temperatures. Thermal background noise, which is much higher in the mid infrared when compared to shorter IR wavelengths, impacts the sensitivity and consequently the BER performance.

Current systems rely on mature semiconductor laser technology and devices manufactured to support the fiber-optic cable industry. Can the components be obtained cheaply? Does the technology even exist to use other wavelengths? The engineering challenge is to use the correct combination of existing and novel technologies to achieve innovation at a reasonable price.

Summary

This chapter looked at a number of the issues that must be considered for a full understanding of the real-world performance of FSO. Weather, link distance, scattering, absorption, turbulence, misaiming, laser wavelength, and data rates all have an impact and must be factored into either a custom calculated link budget or a manufacturer's distance rating.

As future generations of FSO equipment begin to emerge, exciting prospects for even more effective mitigation of issues impacting FSO performance will come about. One example might be low-cost active aiming designs that will allow for low-beam divergence approaching .25 mrad or even less to improve foul weather performance, extend distances up to 10 km or more, and self-aim initially, which would eliminate a tricky part of the installation process. Current 3 and 4 laser systems might be replaced with 8 and 12 laser systems. These potential future enhancements will likely accelerate the deployment of FSO in metropolitan area networks, the subject of our next chapter.

Having discussed some fundamentals of infrared radiation propagation through the atmosphere, this book will start to look at networking issues in the following chapters.

Sources

[1] H. Weichel, *Laser Beam Propagation in the Atmosphere*, pp. 12–66, SPIE Optical Engineering Press, Bellingham, WA, 1990.

[2] R. D. Hudson, Jr., *Infrared System Engineering*, p. 115, Wiley & Sons, 1969.

[3] G. P. Anderson, A. Berk, P. K. Acharya, M. W. Matthew, L. S. Bernstein, J.H. Chetwynd, H. Dothe, S. M. Adler-Golden, A. J. Ratkowski, G. W. Felde, J. A. Gardner, M. L. Hoke, S. C. Richtsmeier, B. Pukall, J. Mello and L. S. Jeong. *MODTRAN4: Radiative Transfer Modeling for Remote Sensing, In Algorithms for Multispectral, Hyperspectral, and Ultraspectral Imagery VI.* Sylvia S. Chen, Michael R. Descour, Editors, Proceedings of SPIE Vol. 4049, pp. 176–183, 2000.

[4] G. L. Stephens, *Remote Sensing of the Lower Atmosphere*, pp. 20–22, Oxford University Press, 1994.

[5] V. E. Zuev, *Laser Beams in the Atmosphere*, p. 215, Consultants Bureau, New York, 1988.

3

FACTORS
AFFECTING FSO

Integration of FSO in Optical Networks

IN THIS CHAPTER

The advent of optical technology has changed the dynamics of global networks. As more and more optical elements are deployed, it is becoming clear that optical networking will play a major role in future global communications. The technological superiority of optical communications—the process of sending voice, data, and video over light signals—compared to alternative modes of communication, such as sending electrical signals over copper and cable, can be substantial. The economics of optical networking has already resulted in its total dominance of the long-distance segment of the global terrestrial and undersea communications network. With exponential advancement of optical technology, its value proposition is becoming compelling even for shorter distance applications. The optical network is steadily advancing through the metropolis toward the end user. This chapter takes a fairly granular look at the drivers of optical technology deployment and the developments in the technology that are making it increasingly attractive. It also predicts some outcomes for current optical technology developments, including free-space optics.

The Optical Networking Revolution

Optical technology has brought a revolution to modern-day networks. Consider the fact that in a rather short span of three years, a technology called Dense Wavelength Division Multiplexing (DWDM) has increased the capacity of a single strand of fiber, thinner than the human hair, over 50-fold. Today, DWDM makes it possible to carry the entire globe's communications traffic—including every telephone call being made, every e-mail being sent, every Web page being downloaded, by every person in the world—over a single strand of fiber. Consider that today, powerful laser devices and optical amplifiers enable a light signal to carry communications traffic for more than 300 miles without the aid of electronics. If that were not impressive enough, optical technologies such as chirped lasers, Raman amplifiers, solitons, and Forward Error Correction, are poised to extend the span of light signals to several thousand miles, making many expensive electronics obsolete in tomorrow's global network.

This is significant because it will bring down the cost of the network by an order of magnitude, giving the next-generation network owners a huge cost advantage over legacy carriers. Consider that with the deployment of Optical Cross Connects and Optical Add-Drop Multiplexers, wavelengths of light can be added to or dropped from the optical network en route between the main hubs. This connectivity will soon allow end users, beginning with large enterprises, to access wavelengths of light from the optical network directly. This not only unclogs a critical bottleneck between the network backbone and the end user, but also opens up the possibility of a "pull" network—a network in which users can demand and receive practically unlimited capacity. Usage will no longer be constrained by fixed capacity pipes that need to be provisioned months in advance. Each of these advances in optical networking—giant strides in their own right—add up to create a powerful and versatile communications network, the likes of which have so far been only imagined.

The modern day Internet revolution is being fueled by the optical networking revolution. It is as if optical networking were accelerating the connectivity needs of an exponentially growing user base. Few would argue against the notion that the Internet and wireless have been the most influential forces shaping the communications network today, and that they will continue to play a pivotal role in defining the communications network of tomorrow. The Internet has grabbed the imagination and the wallet of both retail and business customers by offering a new portal for delivering services—mail, customer service, order processing, entertainment—of every shade and color. The wireless revolution, on the other hand, is enjoying unprecedented customer penetration by offering an untethered portal. In the next wave, wireless and Internet are poised to converge, as data—e-mails, documents, news clips, stock prices, graphics, Web pages, and even streaming video—are delivered over the airwaves to untethered wireless devices.

However, in all the discussion of the Internet and wireless revolutions, one point is often lost: Underlying both communications technologies is a physical infrastructure. That infrastructure increasingly is powered by light, and as more and more users move to the untethered world, light will be the only enabling bandwidth medium. With the advent and adoption of free-space optics, the same light that is now confined by glass will be able to deliver the optical capacity through air, enabling completely optical and untethered connectivity.

The Internet and optical networking have a symbiotic relationship. If the Internet is like a tree, spreading its branches into every conceivable nook and cranny of the global economy, then optical networking represents its roots. Every Web search you conduct, every online auction in which you participate, every book you order, every possible interaction you have with the Internet, results in the exchange of data—data that must be carried by the communications infrastructure. The phenomenal rate of adoption of the Internet has, therefore, been accompanied by a concomitant explosion in data traffic—a 500% increase per year! Yesterday's telecommunications infrastructure, customized for voice, simply cannot handle this data load, and left to its own devices, would have long choked the flow of the Internet to a trickle.

Optical networking, with its tremendous capacity, flexibility of provisioning, and ability to handle data/voice/video with equanimity, is enabling the global communications network to cope with the demands of the Internet. The role that optical networking technology plays today as an "enabler of the Internet" will become even more critical tomorrow. The next-generation Internet will serve as a portal to bandwidth hog applications—3D graphics, real-time streaming audio and video, telemedicine—and cannot really take off without the power and versatility of optical networking in its physical layer.

Although wireless offers an incredibly convenient, untethered portal for delivering communications services, it, too, is rooted in a physical infrastructure. When you speak into your wireless phone, your voice does not reach its destination directly over the ether—even if the person you are calling on his wireless telephone is standing right next to you. Your call is first transmitted to a base station—the one with the giant antenna—and from there it is backhauled to the nearest central office of the wireless service provider. As the number of wireless users has increased dramatically, the capacity of the backhaul pipes has become increasingly congested. Optical technology is stepping in to clear up the pipes; however, the capacity requirements of today's voice-driven wireless network pale in comparison to the capacity requirements of tomorrow's wireless Internet. Without optical networking, the wireless Internet—a portal delivering bandwidth hog applications such as Web page viewing, streaming video, and document transfer over the airwaves—cannot take off.

An important concept to remember is that although optical networking offers such a promise, it is not ubiquitous today. It is limited by the physical infrastructure—in this case fiber—and its reach. Fiber is not available everywhere, which makes it impossible to have optical reach everywhere. Free-space optics extends the reach of optical networks. The promise of the all-optical network seems more and more achievable.

However, understanding optical networking is no easy task. No other industry exists where the cutting edge of deployed technology is so sharp. The technology adoption rate in this sector is so fast that "esoteric" optical technologies, which some colleagues were toiling to understand in Physics labs only a few years ago, are being rolled out as commercial systems today. Even for the technology-trained individual, the industry jargon—Dense Wavelength Division Multiplexing (DWDM), Erbium Doped Fiber Amplifier (EDFA), Distributed Bragg Reflector (DBR) laser, OC-192, and now FSO to name a few—can be intimidating and confusing. Moreover, the description of the technology is often buried in scientific journals and academic research papers. And yet, a basic understanding of optical networking technology is critical to understanding the global communications industry today and how it will evolve tomorrow.

Benefits of Next-Generation Optical Networking

The Plain Old Telephone System (POTS) has served the communications service industry well over the past few decades—its reliability is legendary, its underlying engineering is sound, and it has adapted to incremental increases in communications traffic with the use of older generation optical technologies. Why change a good thing by implementing a next-generation, revolutionary optical technology? It was done not to fix a problem, but rather to add to capabilities.

For a new technology to usurp an older, time-tested technology, the newer version must satisfy at least one of two criteria:

- It must enable functions or services that are not possible with the older technology but are demanded by newer circumstances.
- It must accomplish the tasks performed by the older technology more efficiently in terms of cost or performance.

Next-generation optical networking does both—it enables services and functionality that cannot be matched by any of the older generation communications technologies, and it does so at a lower cost. But before going into more details, this chapter will take a quick look at the older generation optical networks—based on SONET—to understand what the new generation technology brings to the table.

The Recent Past: SONET/SDH

Ever had the opportunity to listen in on two communications network experts conversing? Then chances are you have heard the term "OC-48" or "OC-192" thrown around frequently. They are part of the vocabulary of the SONET hierarchy, a set of standards that define the data-handling capacity of communications networks.

SONET is an acronym for Synchronous Optical NETwork, whereas SDH is an acronym for Synchronous Digital Hierarchy. SONET and SDH are slightly different versions of the same standard, with SONET customized for the North American networks, and SDH customized to accommodate the slight differences in the European and Japanese communications networks. SONET/SDH are basically multiplexing schemes—aggregating communications traffic into higher and higher data rates—allowing the network to be used more efficiently. SONET was developed to fix problems with the earlier PDH (Plesiochronous Digital Hierarchy) standards and define even higher data rates starting with OC-1 (810 simultaneous telephone circuits) and up to OC-192 (10 Gigabits per second of capacity). Table 4.1 reviews some capacity definitions.

TABLE 4.1 PDH and SONET Data Rates and Comparisons

Standard	Level Name	Data Rate (Mbps)	Equivalent Voice Circuits	Relative Capacity
PDH	DS-0	0.064	1	Base
	DS-1 (T-1)	1.544	24	24 times DS-0
	DS-2 (T-2)	6.312	96	4 times DS-1
	DS-3 (T-3)	44.736	672	7 times DS-2
	DS-4 (T-4)	274.176	4,032	6 times DS-3
SONET	OC-1	51.840	810	Base
	OC-3	155.520	2,430	3 times OC-1
	OC-12	622.080	9,720	4 times OC-3
	OC-24	1,244.320	19,440	2 times OC-12
	OC-48	2,488.320	38,880	2 times OC-24
	OC-192	9,953.280	155,520	4 times OC-48

4

INTEGRATION OF FSO IN OPTICAL NETWORKS

Only yesterday, requests for a DS-3 or OC-3 were the domain of Fortune 500 enterprises and carriers. Today, it is not uncommon for an e-Commerce company of 20 employees to order such capacities.

First-Generation Limitations

Optical technology is not new to the communications network; carriers have deployed fiber and optics in their networks since the 1980s. These optical networks, which are referred to as "first-generation," have ring-like topologies and are based on the SONET standards. In communications lingo, they are commonly referred to as *SONET rings* (see Figure 4.1). The topology of SONET rings incorporates point-to-point links between nodes; nodes allow traffic to be added or dropped at major hubs, typically population centers.

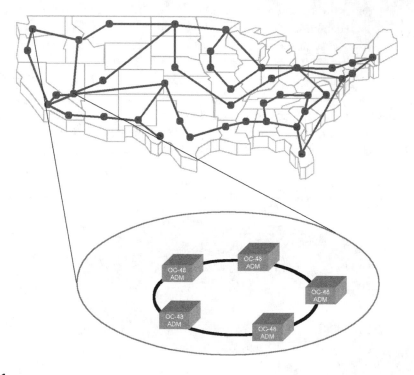

FIGURE 4.1

SONET ring network with Add-Drop Multiplexer (ADM) transport equipment. Source: Hasan Imam, Thomas Weisel Partners, formerly at DLJ.

These networks are powered by equipment called *SONET Add-Drop Multiplexers (ADM)*. Inside the SONET ADM boxes are electronic chips that take lower-speed data streams, such as an OC-3, or an OC-48 signal, and multiplex/aggregate them up to a higher-speed data stream, such as an OC-192. The multiplexed signal is then converted into an optical signal and put on the fiber.

The problem with SONET is that the electronic hardware in the SONET ADM boxes is both specific to data rate and to the protocol of the traffic (see Figure 4.2). This makes the SONET architecture quite complex, rigid, and expensive. Additionally, the footprint (or physical space taken up) of the equipment is significant—and real estate in carrier central office facilities is an expensive commodity. Furthermore, each SONET box output stream is peaked out at 10 Gigabits (Gb) per second. To get more bandwidth, perhaps 20 Gigabits per second (Gbps), two fiber strands would have to be lit up by two SONET boxes—practically doubling the cost.

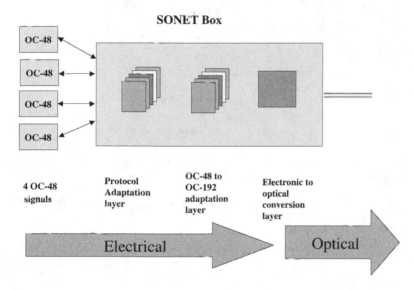

FIGURE 4.2
The guts of a SONET box with its several layers of complexity. Source: Hasan Imam, Thomas Weisel Partners, formerly at DLJ.

The Second-Generation Revolution

A compelling argument for the need for an evolution from the older-generation network to the next-generation optical networking is pictorially presented in Figure 4.3. Consider the number of layers that comprise the older-generation optical network. Sandwiching the optical layer on either side are the SONET Add-Drop Multiplexer boxes. The SONET ADM box, in turn, consists of two elements:

- The SONET Mux/Demux boxes that convert the electrical signal into light, and vice versa
- The Digital Cross Connects, which are massive switch fabrics that electronically add-drop communication signals.

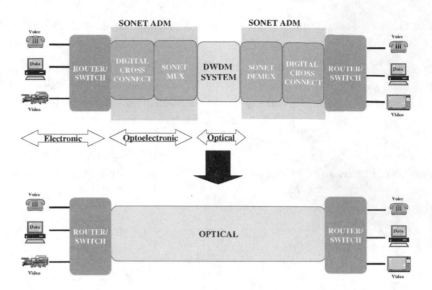

FIGURE 4.3

Simplified next-generation optical networking. Source: Hasan Imam, Thomas Weisel Partners, formerly at DLJ.

Next-generation optical networking aims to eliminate the SONET ADMs altogether, and simplify the communications network significantly.

Next-generation optical networking technology incorporates three revolutionary network elements:

- Optical amplifiers
- DWDM
- Optical switching

Together, these technologies create an extremely versatile, and powerful, communications network. Elimination of the SONET layer alone provides significant advantages. However, that is not the sole benefit presented by the optical network. These additional benefits will be discussed next.

Addressing Bandwidth Demand

Next-generation optical networking technology is the only solution to the explosive demand for bandwidth. No other technology can keep pace. Driven by the Internet, communications traffic is at least doubling every year. Carriers are scrambling to meet the demand by building new networks and upgrading older ones. However, the only technology that allows carriers to keep ahead of the demand curve is optical networking. DWDM technology, which is explained in greater detail later in this chapter, has increased the capacity of a single strand of fiber by 50 times over the past three years. Table 4.2 compares capacity between optical and electronic technologies.

TABLE 4.2 Relative Capacities for Optical and Electronic Technologies

Technology	Medium	Maximum Bandwidth#	Maximum Distance (Without Amplification)
DSL	Copper twisted pair		
	H (high) DSL	~1.5 Mbps*	9,000 ft
	ADSL	~8 Mbps	18,000 ft
	VDSL	~52 Mbps	1,000 ft
Cable Broadband	Coaxial cable	60 Mbps	12,000 ft
Optical DWDM	Fiber	1.6 Tbps**	60 miles

*Mbps= Megabits per second; **Tbps=Terabits per second; #Currently available in deployed or prototype, not a theoretical limit. Source: DLJ*

Providing Cost-Efficient Operation

Among available communications technologies, optical networking provides the most cost-efficient transmission of bandwidth. Next-generation optical networking, with the use of DWDM technology and optical amplifiers, is bringing the cost of transmission down dramatically—both for new network builds and for upgrades (see Figure 4.4). In fact, of all the communications technologies available, optical networking has the lowest cost structure in terms of bandwidth transported per unit distance. Carrying a Gigabit per second (a billion bits per second) worth of information over a kilometer is cheapest on optical networks.

It is the cost efficiency on a per bandwidth–distance basis that has made optical networking the technology of choice for every new long-distance network build in recent years. This includes all deployed *and* planned network builds by Qwest, Williams, Level 3, Enron Communications, AT&T, 360 Networks, and other long-distance carriers.

Optical networking provides the most cost-efficient network upgrade in capacity exhaust situations. Due to the explosive growth of the Internet, fiber exhaust—a situation in which the existing fiber infrastructure can no longer handle the communications traffic load—is quite common, especially in metropolitan areas. Depending on whether a dark fiber (fiber that is in the conduit but has not been hooked up to optics yet) is available, a network carrier typically has four capacity expansion choices:

- Upgrade to speedier electronics.
- Light up dark fiber, if available.
- Lay more fiber, if dark fiber is unavailable.
- Light up the existing fiber with new optical networking gear.

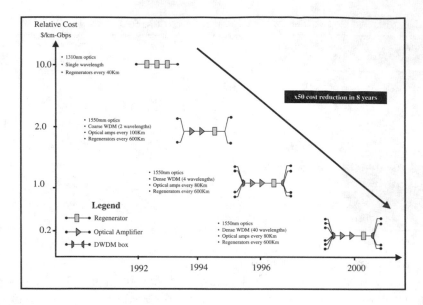

FIGURE 4.4

Relative cost reduction of carrying information using new-generation optical technology. Source: Hasan Imam, Thomas Weisel Partners, formerly at DLJ.

You can do a quick cost-benefit analysis of these alternatives. In doing the math, do not split hairs on the numbers; the goal is to get an order of magnitude estimate for the cost of each of the three alternatives. Also, to keep variations in network architecture from overly complicating the example, use the simple geometry of a 10-mile long point-to-point network in a fiber exhaust situation. Assume that the network has a capacity of 2.5 Gbps or, in Synchronous Optical Networking (SONET) terminology, it is an OC-48 network. The carrier decides that it needs to expand the capacity of its network by a factor of four—that is, to a 10 Gbps or an OC-192 network.

The options include the following:

- It is possible to replace the OC-48 SONET box at the two ends of the network with four times speedier electronics, or OC-192 SONET boxes. The cost of the OC-48 to OC-192 SONET upgrade equipment is $200,000.

- Suppose that unlit, or dark, fiber is available in the conduits. To increase the capacity to 10 Gbps, one option is to light up three additional strands of dark fiber with OC-48 SONET boxes. At a cost of $70,000 per OC-48 SONET box, this cost total equals $210,000.

- The third option requires fiber build. This includes the cost of fiber material ($500/km per fiber strand), fiber sheath ($10,000/km), fiber innerduct ($5,000/km), and fiber conduit structure expansion ($350,000/km)—for a total cost of about $3.7 million! This is without considering the cost of securing rights-of-way and regulatory approval, which can be substantial in metropolitan areas.

- The fourth option is to use DWDM technology, whereby the OC-48 SONET boxes can be ripped out and replaced by a 4-wavelength DWDM system, with each wavelength carrying 2.5 Gbps. The cost for such a system is on the order of $30,000/wavelength, for a total cost of $120,000.

If the distances were shorter, a fifth option would be to use FSO.

Optical networking using DWDM is clearly the least costly alternative. It is about 30 times less expensive than the fiber build option, and about 67% less expensive than the SONET alternatives. It should also be pointed out that if you wanted to expand the capacity beyond 10 Gbps, then option 1 alone would not suffice because commercially available electronics today do not allow data rates greater than 10 Gbps.

Simplifying Network Switching

Optical switching technology can simplify and reduce the cost of network switching: Today, the switching function in optical network nodes is performed by optoelectronic switching fabrics. The optoelectronic switching fabric makes it necessary for the communications signal to go through a costly regeneration process—a process during which an optical signal is converted to an electrical one, electronically processed and enhanced, and finally converted back to an optical signal. This process is also known as an *O-E-O conversion*. In multiwavelength networks, which carry signals on multiple wavelengths/colors of light, regeneration requires separate transmitter and receiver components for every wavelength. This adds to the cost and complexity of the node.

For "express" traffic, a signal just passing through the node (this is true for 30–80% of the traffic passing through a typical node), regeneration might be unnecessary. Switching these light signals in the optical domain, thereby bypassing the regeneration process, can lead to substantial cost savings as well as eliminate some complex control electronics. However, it might not be desirable to eliminate completely the electronics in the switching layer. Rather, the goal is to minimize the role of the electronics to signal monitoring and perform the switching function in the optical domain.

Providing Transparency

Next-generation optical networking technology is content, protocol, and bit-rate independent. Unlike electrical transmission, optical networking is transparent to signal content, protocol, and data rates. After the interface equipment puts an electrical signal on light, the optical network does not care about what the content of the signal is (whether the signal is voice/video/data), what the signal protocol is (whether the underlying signal is in IP, ATM, or Gigabit Ethernet format), or what the data rate is (light can carry a 2.5 Gbps signal just as easily as a 10 Gbps signal). This is an enormous benefit of optical networking because it allows one optical network to haul signals of all different protocols, content, and data rates (see Figure 4.5). Only the electronic interface card at the edge of the network needs to be changed.

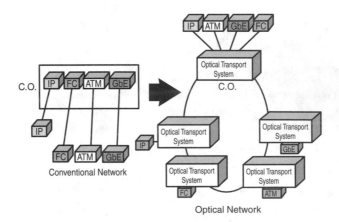

FIGURE 4.5

A conventional network architecture requires separate networks for different protocols, whereas an optical network can handle all four protocols with ease. Source: Hasan Imam, Thomas Weisel Partners, formerly at DLJ.

Scalability

Next-generation optical networks are easily scalable: In a SONET network, an upgrade to greater bandwidth requires every SONET box along the network to be ripped out and replaced with higher data rate SONET boxes. Consider a 4-node OC-48 network being upgraded to OC-192. This involves replacing each of the four OC-48 SONET boxes with four OC-192 boxes. Contrast this to a DWDM optical network; in this case, the optical transport or DWDM units at the nodes do not need to be replaced. All that is required is to replace the electronic interface cards from 2.5 Gbps to the higher data rate 10 Gbps card. Moreover, the SONET architecture cannot scale beyond the OC-192 or 10 Gbps data rate because that is the limit for commercially available SONET systems today. On the other hand, depending on the number of channels on the DWDM system, it can be easily scaled to hundreds of Gigabits per second or even up to Terabits.

Enhancing Reliability

Optical components are inherently more reliable than their electronic counterparts, which decreases the point-of-failures in the network. Electrical transmission generates heat, and heat is the primary cause of failure in communications devices. Optical communication, in contrast, is frictionless; it does not generate heat except when the signal is converted back into electronic format. Using optical components and devices to reduce electronic layers in the network, next-generation optical networking technology is making the communications network more efficient and less prone to failure.

From a network owner's perspective, the following equation holds: For every $35 spent in capital expenditures to build the network, $65 is spent in ongoing operating and maintenance costs. These costs include power, spare inventory, real estate, environmental control, and so on. Optical technology can reduce this cost significantly because optical components and modules need less power than their electronic counterparts. This saves cost, in terms of both power and cooling. Optical components are inherently more reliable; this makes failures less frequent and reduces repair costs. Optical networking equipment typically has a lower footprint than its electronic counterparts; this saves cost in terms of real estate, which is a valuable commodity in many carrier central offices.

Benefits Summary

In summary, optical networking technology has many advantages, some of which include the following:

- It is transparent to the protocol and handles data/voice/video with equal ease.
- It enables enormous capacity expansion.
- It is scalable (future-proof).
- It can be deployed rapidly.
- It can be upgraded with minimal service interruption.
- It is dynamically provisionable.
- It allows enormous bandwidth to be managed with relative ease.
- It allows protection and restoration, which is critical because one fiber carries enormous amounts of data.
- It reduces the footprint (that is, the amount of occupied real-estate) of transport equipment.
- It allows protection and restoration, which is critical because one fiber carries enormous amounts of data.
- It enables new revenue-generating services, such as telemedicine, remote mirroring of data, and so on.
- It does all of the above on a lowest-cost-per-bandwidth basis.

Classifying the Global Optical Network

The global optical network provides connectivity between end users. Using light pulses as units of transmission, the optical network interconnects communications devices—telephones, computers, and videoconferencing units—allowing physically separated users to communicate with one another. When you make a phone call or click on a Web page, the signal from your telephone or the data from your computer can travel to its destination and back in the form of light pulses. Depending on the destination of your call or the address of the Web site you typed, the signal might travel only to your next door neighbor's phone, or it might travel thousands of miles to the other side of the globe. Along the way, the signal can pass over many geographical boundaries, including a local neighborhood, a metropolis, a continent, or an ocean.

The global communications network can be classified and segmented in many ways. One way is by the *protocol*, or format, of the signal that is being carried in the network. A network carrying Internet Protocol (IP) traffic would be called an IP network, whereas another, carrying Asynchronous Transfer Mode (ATM) traffic, would be classified as an ATM network. A second classification scheme involves traffic *content*—voice, data, or video. Yet a third classification scheme is based on geographic span—long-distance, metropolitan, and local access.

For optical networking, the classification scheme that makes the most sense is by geographical span. This is because an optical network does not care what the content of the signal is—it can carry voice, video, or data with equanimity. Nor is the format of the signal an issue; IP, ATM, Frame Relay, or Gigabit Ethernet can each ride on separate wavelengths of light, and simultaneously, over the same optical network. Classifying an optical network in terms of either signal content or format is not as relevant as in the case of legacy networks. What does make sense, however, is to segment the optical network in terms of its geographical span. The capacity, performance, and cost requirements for the different spans of the network can be very different. By this measure, the optical network can be classified into three main segments:

- Long haul: The long-haul, or long-distance, part of the communications network covers a large geographical area, typically over 100 miles, and connects major city centers or traffic hubs. Depending on whether the network stretches over land or is submerged underwater, the long haul can be terrestrial or undersea.

- Metropolitan area network: As its name suggests, the metropolitan area network (MAN), or metropolitan for short, typically refers to the network segment that spans a metropolitan area such as New York or Atlanta. A metropolitan network can stretch anywhere from 10–200 miles. The metropolitan can be further divided into two metropolitan core/backbones, which connect major traffic hubs or central offices (CO) of telephone companies, and metropolitan access, which connects the central offices of the telephone company to customer premises.

- Local area network: Also referred to as access and edge, the local area network (LAN) typically spans short distances—on the order of a few hundred feet to 10 miles. This might be the cable or telephone network in your neighborhood, a network connecting different buildings on a university campus, or even a network inside a large office or apartment building.

From here, this chapter will discuss the role that optical networking plays in each of these segments.

Long-Haul Optical Networking

The long-haul segment of the optical network covers a large geographical area, typically over 100 miles, and connects major city centers or traffic hubs. The long haul, in turn, can be terrestrial or undersea, depending on whether the network stretches over land or is submerged underwater. A long-haul example, the network that connects New York City to Seattle, is a terrestrial long-haul network, whereas the one connecting New York City to London is an undersea or submarine network.

In the long-haul segment of the network, the name of the game is *capacity* and *lowest cost per unit of data transmitted*. These two requirements are associated with, respectively, traffic aggregation and capital cost of deploying the long-haul trunks.

The long-haul pipe must have high capacity because large amounts of traffic are aggregated on it for transport from major population centers. For example, the long-haul network between New York City and Los Angeles needs to have enough capacity to haul all of the communications traffic between the two cities—every telephone call, every Web page download, every e-mail and videoconference between the residents of the two cities. Moreover, this long-distance trunk might also need to carry traffic destined for intermediate cities, such as Boise.

The cost of obtaining rights of passage over hundreds of miles of the network span, and the deployment and maintenance of the physical infrastructure, makes the long-haul network capital intensive. Before deregulation, carriers could simply recoup this investment by passing on the cost to consumers in the form of high long-distance rates. After deregulation, competition is forcing carriers to become more cost conscious. Furthermore, the traffic in the long-haul network is increasingly data, for which the revenue equation is different than for voice traffic. Thus, carriers are interested in deploying technology in the long haul that offers the lowest cost per unit of data hauled.

Consider the following: Using DWDM technology that is commercially available today, a carrier can load as much as a Terabit (2 trillion units of digital information) of data on a single strand of fiber. No other technology comes even close to offering this capacity of bandwidth.

Although optical networking offers sheer raw bandwidth through DWDM and reduces the cost per unit of data carried that way, DWDM is not the only way that optics brings down the cost of transmission. Optical amplifiers also play a significant role. Before optical amplifiers, the network required an expensive electronic box called the regenerator every 40–100 kilometers depending on the type of fiber and light source used (see Figure 4.6). With the advent of the Erbium Doped Fiber Amplifier (EDFA), the regenerator is needed only once every 600 kilometers, with the optical amplifier replacing the regenerator in the intermediate span (see Figure 4.7).

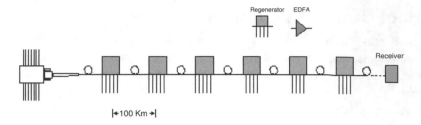

FIGURE 4.6

Before EDFA, expensive regenerators were needed every 100 kilometers. Source: Hasan Imam, Thomas Weisel Partners, formerly at DLJ.

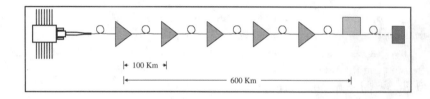

FIGURE 4.7

After EDFA, regenerators are needed only every 600 kilometers. Source: Hasan Imam, Thomas Weisel Partners, formerly at DLJ.

The combination of high capacity and the low cost per unit of data transmitted over a unit distance makes optical networking the undisputed technology of choice for the long haul. Playing to this market opportunity, a number of vendors offer optical networking systems to the long-haul carrier market.

The cycle of greenfield (newly built) network builds by new carriers that are taking advantage of global telecommunication deregulation, and the upgrade by incumbent carriers in response, continues to accelerate demand for long-haul optical networking systems.

An emerging growth area is ultra long-haul optical networking, whereby a communications signal can travel up to 5,000 Kilometers without requiring any electronics. This will bring down the cost of long-haul transport substantially and will be adopted in greenfield networks of the future.

Metropolitan Area Networks

As its name suggests, the metropolitan area network (MAN), or *metro* for short, typically refers to the network segment that spans a metropolitan area such as New York or Atlanta. A metropolitan network can stretch anywhere from 10–200 miles. A MAN can be further divided into two MAN components, the core/backbone, which connect major traffic hubs or central offices (COs) of telephone companies, and MAN access, which connects the central offices of the telephone company to nearby customer premises (see Figure 4.8).

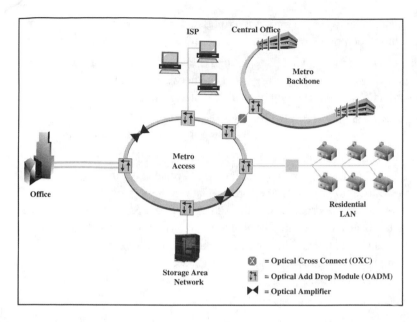

FIGURE 4.8

The metropolitan area network (MAN). Source: Hasan Imam, Thomas Weisel Partners, formerly at DLJ.

Optical networking has a foothold in the metropolitan market today through the fiber infrastructure that is already in the ground. However, the fiber is powered by older-generation optical equipment—SONET boxes. The SONET layer is rather inflexible and cannot be scaled gracefully to accommodate the explosion in data traffic. SONET is giving way to next-generation optical networking in the metropolitan networks.

When you take a closer look at the market for the MAN, you begin to realize that this segment of the network market is rather complicated. The MAN has a variety of customers with a variety of needs. The MAN aggregates traffic that comes in different formats and bit rates. It connects equipment boxes that are often incompatible. For example, Gigabit Ethernet cannot be plugged into a standard SONET box; therefore, it does not enjoy the protection afforded in the SONET layer.

If you were to compile a list of the demands and needs of the various groups of customers—Internet service providers, storage area network vendors, telecommunications service providers, enterprises—to which a metropolitan network must cater, the list does not just include capacity or lowest cost. Rather, it is a long list of demanding features. Any networking platform that will be adopted widely by all these groups must have the following features:

- Increased capacity: Handling the exponentially increasing traffic in the metropolitan core is a primary consideration for these networks.

- Rapid deployment: The time-to-market pressure on competitive carriers is high; delay in expanding or deploying capacity means lost revenue streams and lost opportunities.

- Scalable: The traffic demand patterns are unpredictable, except for the fact that they are increasing. Carriers need to future-proof the network, which means building a system that is rapidly scalable and without service interruptions.

- Service transparency: The platform must be able to handle a variety of protocols, including ATM, Frame Relay, IP, Gigabit Ethernet, Fiber Channel, and so on.

- Protection and restoration: With one fiber carrying millions of phone calls, and with the advent of online trading, online banking, and remote caching of time-sensitive data, it becomes critical to protect and restore the data in the metropolitan network. Protection is important for services such as Gigabit Ethernet, where SONET is bypassed and the protection that is given to data by the SONET layer is no longer available. Furthermore, a metropolitan networking platform must be able to restore services within 30 milliseconds—the requirement by telecommunication service providers—if it is to get Inter Office facility business.

- Manageable: Every customer says, "Can't manage it, won't deploy it—period." Just raw capacity is not enough; the platform must provide the means to manage the capacity.

- End-to-end dynamic provisioning: Metropolitan traffic is unpredictable. A bottleneck might appear anywhere, anytime. Therefore, a metropolitan platform must be capable of end-to-end provisioning; it must be able to optimize the resources of the network anywhere, anytime.

- Lowest cost: The universal metropolitan platform must be able to provide all these features at the lowest first cost. Competitive carriers require low initial capital cost in service deployment. However, it is important to keep in mind that without the other features, the cheapest platform will not win. One customer said, "Cost is always a factor, but it is not the only factor."

When you compare the needs of the metropolitan customer with what optical networking technology has to offer, as discussed previously, there is a strong fit.

- New services—wavelength for hire: The introduction of WDM networks into metropolitan areas opens up the possibility of offering a *wavelength-on-demand*; that is, provide virtual bandwidth pipes to customers who require large bandwidth. For example, perhaps DLJ needs to videoconference with all its offices at 1 p.m.—a huge bandwidth demand. Williams is providing this service already under its Optical Wave Service. Such leasing is also time efficient; although it will take Williams six months to provision an OC-48 SONET channel on their traditional network, Williams can provide the lambda in six weeks. Moreover, lambda services require less power (one-third of traditional SONET gear) and reduce space requirements by as much as five times.

The metropolitan demand for bandwidth is being driven partly by new bandwidth-intensive services, such as storage area networks (SAN) and Gigabit Ethernet (GigE) Enterprise networks. Some metropolitan optical system vendors will integrate vertically with SAN and GigE hardware vendors.

The time is right for the new generation of optical networking to establish a foothold in the metropolitan market. It is key to have systems built 100% from the ground up for the metropolitan network. Currently, most of the platforms are really extensions, drummed down versions of the vendor's long-haul platform or souped-up versions of the access platform. The key is to understand and implement the needs of metropolitan networks as features in their platform.

Access Networks

So far, this chapter has talked about long-haul and metropolitan segments of the communications network. Now it will focus on the segment of the network that actually touches the end user: Access. Sometimes also called the local area network (LAN), Access refers to the segment of the network with a short span, ranging from a few hundred feet to 10 miles. The cable and telephony network in your neighborhood, the network connecting all the separate buildings on a university campus, and the network inside a large office or apartment building are all examples of an Access network.

The Access network can be further classified into three types:

- Enterprise LANs: These are networks that are owned and operated by the enterprise or Internet service providers (ISPs). They reside within business campuses or within buildings. Voice in such networks is typically handled using X.25 or ATM protocols, whereas data is handled by a variety of protocols such as ATM, Fiber Channel, ESCON, Ethernet, Fast Ethernet, and Gigabit Ethernet. Storage area networks (SAN), networks connecting storage devices to the end user, fit into this category.

- Residential Telephony Networks: These are networks that are owned and operated by the Regional Bell Operating Companies (RBOCs). They bring the telephone connection from the carrier's central office to your home.

- Residential Cable Networks: These are networks that are owned and operated by cable service providers, including one Interexchange Carrier (IXC), AT&T. The cable network brings video programs into your household.

Access remains the final frontier for optical networking. This is the segment where the network finally touches the end user. Therefore, the number of connections that branch from the Access network can be large. For example, a cable network can drop off signals to 5,000 or more households, whereas a large enterprise network can connect tens of thousands of computers. The sheer number of connections makes the Access network the most difficult for optical networking to penetrate.

What are the cost issues that stand in the way of optical networks' push into Access? One misconception about fiber's push into Access has been that the raw fiber is too expensive; in most cases, raw fiber is actually cheaper than copper or coaxial cable, and the proceeds from recycling the stripped copper alone can pay for the fiber. Copper's advantage really involves its installed base—simply put, it is already there. Ripping out the copper and deploying fiber in residential cable and telephony networks can be expensive in terms of construction costs and regulatory approval for digging up the neighborhood streets. Even if the network service providers were willing to swallow this cost, the question then becomes how far to push the fiber into the network. The issue here is the cost of lighting up the fiber.

Take the example of a large enterprise, one with a sound telecom budget, which is considering deploying fiber in its data network all the way to the desktop. The issue is not so much the cost of the fiber, but more the cost of the components that are needed to light up the fiber. For a desktop computer to interface with a fiber, it must have an optical network interface card (NIC). Such a card would need to have an optical transceiver—a light source and a photodetector with the associated electronics—to convert an incoming optical signal into an electrical one that is usable by the computer and also to convert the electrical output of the computer into optical. A 100 Mbps (100BFX) NIC costs about $250, and a Gigabit (1000BSX) between $250–$500. For some applications, this might be too expensive, given the number of computers that might be involved. In addition to the NIC, the enterprise network would have to be equipped with optical signal management and aggregation software and hardware.

Although cost is an important issue, the proliferation of the Internet and bandwidth hog applications are driving the demand for bandwidth at the Access level—and when it comes to high-bandwidth delivery, the cost equation for optical networking starts to look more favorable. With the need for bandwidth, optical networking has begun to infiltrate the local access networks. Fiber is pushing deeper and deeper into both the telephony and cable networks, whereas the advent of Fiber Channel and Gigabit Ethernet is making a case for optical networking to push into the enterprise LAN. Parallel to the bandwidth demand phenomenon, innovation and maturing manufacturing processes are starting to bring down the cost of optical networking.

As the demand for bandwidth increases, and as innovation and maturing manufacturing continues to push the "first cost" (the initial cost for deployment) of optical networking down, the technology will make deep inroads into the Access networks.

Now you will learn in greater detail how optical networking is pushing into the three different networks—the Enterprise LAN, the residential cable network, and the residential telephony network.

Cable Networks

Optical networking has made deep inroads into the cable infrastructure. The initial drive toward more optical in cable network architectures has been reducing costs.

In broadcast type networks such as cable TV, the same signal is dropped into a large number of end user homes. In such applications, the original signal needs to be strong enough to allow many drop-off points—sometimes as many as 500–1,000 users can feed signal off of the same pipe. In previous architectures, the signal strength was provided by a large number of electronic amplifiers. However, using fiber that is powered by optical networking gear—or more specifically, multiwavelength DWDM with optical amplifiers—can provide the large signal strength required in a more cost-efficient manner.

The new and more aggressive push to drive optical networking deeper into the cable networks is the *broadband revolution*, pushing voice/video/data into American households using the cable pipes. Driven by AT&T's new cable strategy, one that is predicated upon the upgrade of cable networks to provide *two-way* voice/video/data over cable to residential users, the CATV network is being upgraded with optical networking moving deeper into the network architecture. The optical networking gear dramatically increases the bandwidth of the network.

Because the cable network is a shared network, this additional bandwidth is essential to prevent signal degradation when everyone gets on the network at the same time. Although a certain amount of this degradation is tolerable when the network is only used to deliver video, such degradation becomes unacceptable when the cable pipes are used for two-way telephony and Internet applications. Thus, a two-way broadband upgrade of the cable infrastructure requires the network to be over-engineered in terms of bandwidth. This is accomplished by using the bandwidth multiplying characteristics of DWDM systems as well as re-engineering the architecture of the CATV network to push optics deeper. One implementation of this approach has been accomplished in AT&T's cable networks by placing DWDM systems with add-drop functionality in "Mux Nodes," and then dropping wavelengths into optical "Mini Nodes." The mini nodes comprise optical transceivers that perform optical-electrical and electrical-optical conversion.

Figure 4.9 is an example of a hybrid fiber-coaxial network. In these networks, the fiber extends all the way to the mini-node; coaxial cable takes the signal the rest of the way to the home.

Residential Telephony

The owners and operators of residential telephony networks, the Regional Bell Operating Companies (RBOCS), have begun deploying optical networking technology in their residential telephony networks. This might surprise some because the RBOCs are not particularly well known for their adoption of revolutionary technology. Because the carriers have been granted regional monopoly over the telephone lines reaching into more than 99% of the households in the United States, they have little incentive to rock the boat. However, the emergence of cable broadband has forced the hands of the RBOCS. As AT&T leads the charge to upgrade the cable pipes into two-way broadband pipes, the RBOCS risk losing their stranglehold on the "last mile" of the communications network. The fear is that as the residential users begin to sub-scribe to broadband cable services to experience the benefits of video-on-demand and fast Internet access, customers will switch their telephone services to the cable operator as well.

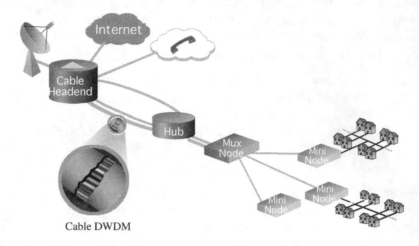

Cable DWDM

FIGURE 4.9

A deep fiber HFC (Hybrid Fiber-Coaxial) cable network. Source: Harmonic Inc. & Hasan Imam, Thomas Weisel Partners, formerly at DLJ.

Fiber is pushing deeper into the residential telephony network in degrees. The degree is characterized by Fiber to the Curb (FTTC) or Fiber to the Home (FTTH). Consider a number of real-life examples:

- Project Pronto: SBC Communications has undertaken a $6-billion effort, called "Project Pronto," to roll out broadband DSL services to more than 80% of the territories under its belt by 2002. As part of the DSL rollout, SBC is pushing fiber and optical networking deep into the loop, the motivation being the large bandwidth. The fiber, powered by optical networking gear, will allow SBC to boost its broadband offering to 6 Mbps from the more common 1.5 Mbps ADSL offerings. The additional bandwidth allows SBC to add video to the list of services it can offer to its customers. (Compressed video takes about 4 Mbps.)

- Bell South's Integrated Fiber in the Loop (IFITL) Initiative: Bell South has undertaken a network rebuild initiative that integrates fiber and copper in its residential telephony network. Starting with more than 200,000 homes in the Miami and Atlanta region, Bell South is pushing fiber as deep as 500 feet of the household. Currently, 95% of Bell South's access network involves fiber within 12,000 feet of residential premises.

- ATM Passive Optical Networking (PON): Another initiative, also undertaken by Bell South, involves pushing Fiber-to-the-Home (FTTH). The experiment involves 400 homes in Atlanta. The project demonstrates that the regional carriers are taking the issue of upgrading their networks with optical technology seriously. The project involves using the Asynchronous Transfer Mode (ATM) protocol to deliver voice, video, and data in an integrated fashion. At the physical layer, the Passive Optical Networking architecture is used. Simply put, this means using only passive optical components, such as splitters and couplers, in the network segment outside the central offices (CO). This saves the cost of building and housing AC powering equipment outside the CO. One drawback of this approach is that the optical transmitter/receiver at the customer premise needs to be powered from the customer's AC outlet. Although not earmarked for mass deployment anytime soon, this approach shows one of the ways that carriers can deploy optical networking all the way to the home in a cost-effective way.

Enterprise Networks

In enterprise networks, the cost-benefit calculation for copper versus fiber so far has been skewed against fiber. Although fiber provides better performance, it costs more to install and light up. However, this is changing with the advent of bandwidth-intensive protocols, such as Gigabit Ethernet and Fiber Channel in enterprise networks. Although Ethernet and Fast Ethernet both involve data rates that are well within the reach of copper, the bandwidth demands of Gigabit Ethernet throw the game open to fiber again. When you add to the bandwidth demand the factor that the cost of optical components for short-distance applications is decreasing, optical networking for the enterprise begins to make sense.

Gigabit Ethernet (GigE)

Gigabit Ethernet (GigE) is a datacentric protocol based on Ethernet and Fast Ethernet, the most popular and widely deployed protocol in computer networks. GigE supports data rates up to 1 Gbps and is compatible with the legacy base of Ethernet and Fast Ethernet. Some of the demand drivers for the higher bandwidth provided by GigE are as follows:

- Internet and corporate intranet: Clearly, the strongest driver of data traffic across the network, the Internet is pushing the demand for Gigabit rate capacity at the Enterprise LAN level. Proliferation of intranets is also adding to the internal demand.

- New applications: Even as the omnipresent Windows Networking operating system requires more and more capacity, multimedia and video applications can demand as much as 1.5 Mbps of continuous bandwidth from LANs.

- Exchange of files electronically: More and more users are exchanging files in the enterprise environment. These files can be anywhere from a few hundred Kilobits per second to several Megabits per second.

Because of the sheer bandwidth that data GigE switches carry, it might be best to connect them directly to optical pipes. Consider the fact that your typical T-1 or T-3 line from the service provider's network to the office basement is not even close to having the capacity to haul GigE

4

traffic into the backbone. At the minimum, you need OC-24 pipes with 1.2 Gigabits (Gb) of capacity, and then you are starting to talk optical. The issue is why to have the GigE connection go through layers of SONET muxing and demuxing. Why not just connect it directly to the optical backbone? This is where next-generation DWDM technology steps in. Putting a GigE or two onto a wavelength of light by interfacing the GigE switch directly to a DWDM box becomes an attractive proposition. As GigE proliferates through the corporate data networks, it opens the doorway to optical networking (see Figure 4.10).

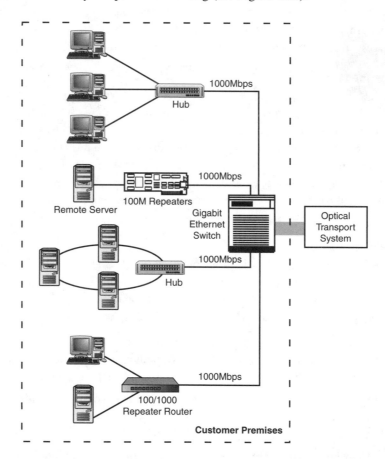

FIGURE 4.10

A Gigabit Ethernet network: opening the enterprise door to optical networking. Source: Hasan Imam, Thomas Weisel Partners, formerly at DLJ.

Fiber Channel (FC) and Storage Area Networks (SAN)

Fiber Channel is a datacentric protocol that supports data rates from 622 Mbps to 4 Gbps. The protocol is designed to optimize fast access in and out of devices and allow fast transport. As a result, the protocol has found a home in Storage Area Network (SAN) applications (see Figure 4.11). The demand for Fiber Channel and SANs is being driven primarily by the following:

- e-Commerce: As data becomes mission critical for many business models (online trading, auction, financial services) the need for real-time archiving and retrieval of data, disaster recovery, and remote mirroring becomes critical.

- New applications: Large e-mail folders, graphics files, audio, and video clips require large storage capacity.

- Data warehousing: More businesses are growing their storage capacity for storing historical data online, analyzing it, and making business decisions from it.

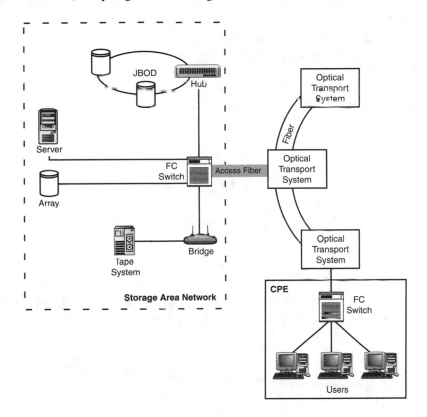

FIGURE 4.11

Optical networking providing connectivity between a remote user and a SAN. Source: Hasan Imam, Thomas Weisel Partners, formerly at DLJ.

Given the data rate requirements of SANs, the only viable method of remotely accessing the SAN is through optical pipes. Even in local SAN applications, accessing bandwidth-intensive files in real time make a case for optical networking in the enterprise.

4

INTEGRATION OF
FSO IN OPTICAL
NETWORKS

Storage area networking is an application designed around providing high-speed native connectivity between storage networks in an enterprise or metropolitan area. The challenge surrounding storage area networking is adapting storage protocols such as Fiber Channel to a standard telecom transport protocol. To date, there has been development in the design of Fiber Channel over ATM, but this is in the hands of a few vendors and is not widely implemented. In addition, this typically creates a network service that does not match the data rate of Fiber Channel, which is most often 1 Gbps. Optical edge gear is not well adapted to support SAN applications unless it can devote an entire wavelength to a connection.

An element of the SAN market is disaster recovery. Disaster recovery is an emerging application for metropolitan DWDM systems in the enterprise market, and represents one of the earliest applications of DWDM to date. As an application of a fiber network, it most often consists of fiber links between corporate data centers and backup facilities, data warehouses, or mirrored data center sites. This is essentially an extension of the SAN market, in which Fiber Channel is currently being utilized to achieve highly scalable, lower-cost storage networks that are used in major corporate data center applications. DWDM disaster recovery systems provide the high-capacity, scalable link between SANs and remote mirroring sites, data warehouses, and other data centers. DWDM offers the advantage over dark fiber through its support of multiple protocols on a single fiber, as well as longer link lengths than Fiber Channel.

At present, this is clearly a limited market, consisting of a small group of large enterprises such as financial institutions, banks, insurance companies, and other major data users. That said, SAN has most often been an application espoused by enterprise DWDM or metropolitan DWDM vendors. Nevertheless, it is a growing segment that will definitely impact the bandwidth demand. [1]

Driving FSO from the Edge

In typical views of metropolitan optical networks, you see two primary segments: metropolitan core and metropolitan access. Significant revenues and innovation have gone into the development of these two segments in anticipation of bandwidth explosion.

A third segment of metropolitan optical networks merits discussion as well: the edge networks. Edge networks, which represent the majority of the end user networks, are being ignored, primarily because they are part of the 95% of commercial buildings that are not connected to fiber. Edge networks represent the driving force behind the growth of optical networks in the metropolitan core and metropolitan access networks. An abundance of underutilized fiber exists in the core of the metropolitan networks because the end user traffic has not reached the core. The answer to this is free-space optics. Free-space optics is enabling the end users to get access to optical connectivity quickly, cost effectively, and reliably.

Storage area networking is one of the drivers contributing to the growth in bandwidth demand at the edge. This section covers several other applications that are driving this growth.

VPN Services

[2] Virtual private networks (VPNs) are driving demand for high-bandwidth Internet access. A VPN is simply an encrypted connection between two or more sites across the Internet using various security strategies capable of constructing a secure "tunnel" through which they can communicate. The justification for a VPN versus point-to-point leased line connections between companies and their branch offices or data centers is that it is much lower in cost.

Broadband Internet Access

ISPs are exerting pressure on carriers to provide them with low-cost broadband access solutions, but they remain frustrated with both the slow response from ILECs and the lack of offerings from competitive carriers.

ISPs are generally stuck offering high-speed access to the Internet backbone via traditional T1 or T3 private lines at a high price compared to projected pricing for ADSL and alternative broadband access services. ISPs are faced with either developing their own broadband access networks at a high cost or buying high-speed access from wholesalers, better known today as *Data CLECs*. As the ISP industry continues to consolidate, these growing organizations will have more capital to directly fund deployment of access facilities rather than simply backbone construction and expansion. In many instances, Internet service providers can evolve into integrated service providers, partnering with competitive access providers to reach end users directly with a full suite of telecom solutions.

Optical edge equipment that addresses carrier needs for incremental growth and scalability will be well positioned in this application space. The choice facing most optical edge vendors is how best to optimize their product while meeting the requirements of carriers and ISPs that primarily serve broadband Internet access. The choices are many, including ATM aggregation optimized for DSL or broadband wireless access, channelized SONET for rapid private-line provisioning, PONs for highly scalable optical access, and metropolitan optical IP platforms for customer-controlled Ethernet-based access. The most promising of these candidates from a broadband access perspective is currently the metropolitan optical IP camp. Optical IP systems, either based on distributed packet switching or Ethernet switching, offer metropolitan carriers an opportunity to operate a managed Ethernet or IP metropolitan network capable of delivering services in increments of 1 Mbps to subscribers over a low-cost Ethernet-based infrastructure.

Systems from vendors in the integrated metropolitan DWDM or the next-generation SONET camp might certainly be in a position to benefit from the deployment of broadband access services, although they will likely find optimal placement closer to the network core. In these implementations, aggregating and grooming DS-3 ATM or 10/100 Ethernet links from building basements, remote pedestals, and first offices will provide more efficient use of network resources. Integrated IP or ATM switching will be required, and it is possible to see the value of integration of subscriber management systems and protocol interworking capabilities to increase the value of these optical edge systems.

Transparent LAN Services

Transparent LAN services (TLAN) have long been regarded as an ideal application in the metropolitan area, providing seamless interconnection of LANs between enterprise and remote locations. TLAN services enabled by Frame Relay and ATM network services were promoted as a key application for ATM VP ring equipment. With the increasing deployment of intranets and extranets, many enterprise customers are opting for network services that provide LAN-speed access to the public Internet rather than dedicated bandwidth between two sites. TLAN services in this scenario can become a subset or application of a VPN. For optical edge network equipment providers, this results in a more simplistic requirement for rapid provisioning of high-speed circuits across the MAN. In this case, one could argue that simple provisioning of access bandwidth is enough, leaving service layer intelligence to IP layer equipment.

Optical Access at Multitenant Buildings

The MTU market equipment is substantial because it is relatively untapped. It is also ultimately limited because of the small number of buildings with multiple tenants that require high-bandwidth services. The addressable market size is roughly 125,000 buildings in the U.S. with 20 or more tenants in an "office" environment.

Today, these tenants typically negotiate directly with various service providers to gain access services. Landlords limit the choice of service providers most often, and they might not grant access rights to their basement or rooftop for the location of access gear. Building LECs are currently striving to overcome these bottlenecks by establishing in-building telecommunications networks that provide a platform for multiple service providers to offer tenants their services through a common platform. This ideally will speed the delivery of services to tenants, reduce the cost of services by creating a more competitive environment, and expand service choice.

The MTU market can be addressed in numerous ways, which creates multiple opportunities for optical edge equipment vendors. Some buildings can benefit most from next-generation SONET. SONET is able to support the transport of standard voice traffic as well as provide distributed cross-connect functionality to provide grooming of voice and data circuits at the network edge. This creates efficient metropolitan network architectures for multiple services. Metropolitan optical IP providers can argue that their low-cost infrastructure is well suited for building access because of its foundation in Ethernet switching, which closely matches the LAN switches inevitably in the building. This allows a broadband local exchange carrier (BLEC) to create pure Ethernet networks that support trunking to metropolitan POPs over a survivable packet ring architecture.

The challenge to optical IP vendors is delivering systems that truly support multiple classes of service. To date, this capability is often promised but rarely delivered. In addition, if a proprietary scheme is used, a BLEC might be forced to build out an entire metropolitan network over dark fiber. This might create cost burdens beyond the limits of a BLEC, which typically spends more than 80% of its budget on in-building network creation.

It is clear that the MTU market will be a driving force and an application that is just around the corner. With more bandwidth pressure, the edge of the network will continue to face the connectivity bottleneck.

Private Line Services

According to recent trends, private lines will account for a larger percent of business telecom spending over the next decade as many applications that were once housed within an organization are moved out into the metropolitan network. Private lines have long been considered the most expensive and bandwidth-inefficient solution for data services, yet their reliability and ubiquitous availability have made them the stalwarts of the Internet era.

Cable Data Transport

Cable TV operators have been examining the broadband data market for nearly a decade, but they have been slow to adopt the necessary infrastructure for reasons of cost, lack of developed standards, and lack of available access equipment. These obstacles were primarily overcome in 1998, and cable TV operators are making important strides in gaining the lead in the broadband access race in the U.S. and the world. At present, the following three key drivers exist in the broadband cable access market in developed economies:

- Multisystem operators (MSOs) have aggressively upgraded their networks to support two-way services using Hybrid Fiber-Coaxial (HFC) architectures, although a great deal of work needs yet to be done.
- The emergence of data-over-cable standards has sparked the cable modem vendor market.
- A serious focus has materialized on enhancing customer marketing programs and customer service.

Cable MSOs have not been as successful in attracting business customers to their data networks, although the @Work service is presently serving more than 2,000 subscribers in North America with high-speed Internet access and remote access services over a cable data infrastructure. The difficulty at present is a perceived lack of quality to support mission-critical data services. In that light, cable MSOs will likely find their greatest success in small and medium-sized businesses seeking only to access the Internet at broadband rates, or provide low-cost telecommuting options to their employees.

The cable industry has long been considered technologically feeble and willing to cut corners to keep profit margins high. This type of operation, or even the perception of it, will be unacceptable in the provision of cable data services, particularly after real competition exists from telcos and satellite operators. The work of CableLabs in North America has improved the industry's image, and the development of true cable data standards will also encourage the use of cable modems beyond the residential sector, although in limited numbers.

The optical layer in a cable television network is often quite different from a traditional telecom network. CATV operators can benefit from the fiber savings at the core of their network, which today is often SONET-based. In addition, CATV operators are beginning to experience the rapid growth and related bandwidth demands that are associated with the success of cable modem services. As these consumer broadband services increase, a scalable core network will be essential for CATV operators to remain competitive with rivals.

WDM technology can also be used to establish a return path on a CATV fiber network and consolidate headends (aggregation hubs for several end users) into "superheadends," greatly reducing network operations and management costs. FSO makes a natural fit into the CATV fiber network and will help service providers to extend the reach of their networks.

DSLAM Aggregation

DSLAM aggregation is commonly cited as one of the most important applications of optical edge networking systems. The reason is fairly straightforward: Today's DSL networks have been quickly assembled using a mixture of co-located DSLAMs and core ATM switches interconnected by traditional SONET transport ADMs. The result is a network in which aggregated DS-3 or OC-3 ATM signals are delivered from a DSLAM to a core ATM switch, which performs the necessary grooming and transport to the appropriate wide area network or Class 5 switch, depending on the services offered by the DSL provider. In these networks, multiple layers of DS-3 access shelves, SONET ADMs, and ATM switching systems are required to support the rapid growth of DSL services.

Operating this network can be quite cumbersome for carriers because of the high volume of traffic that must often be backhauled to core switching facilities, consuming significant bandwidth on metropolitan SONET rings.

Optical edge systems that have been developed with DSLAM aggregation applications in mind tend to employ distributed crossconnect, ATM switching, and SONET transport in a single platform. Through this level of consolidation, a single optical edge system can terminate multiple feeds from DSLAMs, perform grooming down to the DS-0 or DS-1, efficiently fill large optical circuits with service provider–specific traffic, and perform the necessary transport to metropolitan POPs (point of presence). In greenfield networks, this often allows a carrier to build a much more distributed architecture, obviating the need for large, expensive core ATM switches and digital crossconnects.

This will be a recurring mantra in the optical edge networks market: Distribute switching and grooming functionality to eliminate or reduce the requirements for costly centralized switching systems. The benefits to the carrier include improved service velocity, ease of service management, and more efficient use of metropolitan network bandwidth. As carriers become more comfortable with these platforms, they will have opportunities to not only increase operations efficiencies, but to provide new value-added services from them. Many vendors have included the option for time-of-day and day-of-week bandwidth reallocation, customer-controlled network management, and tiered classes of service.

Tiered Optical Bandwidth Services

The addition of an optical layer into the local exchange will provide new methods of protection and restoration. At present, ATM, SONET, and optical layer equipment have their own independent protection mechanisms, designed to work within their own network layer. The difficulty here is that a number of these functions are redundant or inefficient. Optical layer protection mechanisms have the advantage of being protocol and bit-rate independent. Coordinating protection and restoration functions among the network layers is a complex process and is only beginning to be addressed at this time. As in the early stages of the SONET market, hardware is presently outpacing software in the optical networking market. Whereas systems can easily accomplish 16-channel operation today, the software required to manage these wavelengths independently and restore them in complex topologies is not yet available, and is 18 months to two years behind.

The growth of survivable WDM networks will likely parallel the growth of similar SONET architectures. Both DWDM and SONET systems are connection-oriented multiplexed networks. Both employ distributed intelligence (multiplexers) to facilitate protection switching, and a centralized or distributed control mechanism (crossconnects) to facilitate restoration. The potential benefits of optical layer protection and restoration include the following:

- Multigigabit switching and routing in an all-optical domain.
- Reduction in electronic function and cost by migrating protection and restoration to the optical layer.
- The creation of a common survivability platform for all network services, including those without built-in protection capabilities.

Most metropolitan DWDM systems available today depend on the connected SONET equipment to provide protection against node failures or fiber cuts, or they provide optional automatic protection switching. The disadvantages of this approach include the following:

- WDM system duplication, nearly doubling the cost in many cases
- Separate protection system required for each optical channel
- Non-SONET elements not being protected (unless optional APS is employed)

As metropolitan DWDM systems migrate into the access arena, they will be supporting both SONET and native data services, increasing the requirement for protection and restoration in the optical domain. Simple APS is available today on most vendors' equipment, whereas others (Nortel) are beginning to employ electrical crossconnects at the core to provide selectable wavelength protection.

Protection and restoration are not synonymous. In today's network, these functions represent two distinct functions of fiber-optic equipment. Protection refers to the simple, fast (< 50 ms) switching of traffic from one optical route to another predetermined route in the event of a detected failure. SONET equipment performs protection switching today at acceptable rates. Thus, optical layer equipment must perform at least as well to be justified in the network. Protection switching on SONET routes today typically requires 100% excess bandwidth on a given route, which creates additional demand for fiber. Optical line protection will enhance quality of service (QoS) levels for non-SONET traffic—such as ATM and IP—by providing faster restoration than possible in those protocols.

Restoration is a secondary mechanism that can be much slower than protection because it determines routes on-the-fly as nodes fail or become saturated. In optical networks, restoration will be performed by optical crossconnects, most often in mesh topologies. Crossconnects will have the embedded intelligence to select available paths on the network to route wavelengths or entire fibers around saturated or failed nodes. This can lead to more efficient and cost-effective networks as the need for SONET equipment diminishes. Optical layer restoration will be needed for events such as optical amplifier failures, fiber cuts, transponder faults, and SONET LT protection. This will, however, require sophisticated software to compute the efficient alternative routes.

Eventually, restoration will evolve to full wavelength restoration, in which each wavelength will be able to be restored separately. This will require wavelength translation in most cases, but adds the benefit of the most efficient method of utilization of fiber resources. This capability is being realized at the core of networks and not in the metropolitan area at this time.

As noted, most tiered optical bandwidth services proposed today tend to be associated with long-haul network operators. These services usually come in the form of leased OC-n circuits across the wide area and are often wavelength services. In the long-haul network, the deployment of optical switching systems enables this capability, whereas in the metropolitan network, optical switching systems or optical edge systems can provide this capability either at the optical layer (for tiered wave services) or at Layer 1 and 2 for tiered leased line or data services. This kind of flexibility will be appealing to metropolitan carriers that serve ASPs and broadband ISPs because each benefit from the flexible pricing that is associated with tiered bandwidth services and the high degree of customer network management.

Wavelength on Demand

Wavelength on demand is probably the "hottest" MAN service offering from the new breed of metropolitan area carriers with new names like Yipes and Telseon leading the way. The allure is not only cost oriented, which is the intended benefit. The allure can be compared to the exuberance felt by users a couple decades ago as they abandoned the mainframe in favor of doing spreadsheets on their own PC. It definitely connotes "power to the people."

In the long-haul world, wavelength on demand is most often found in the literature of national wholesale network operators. In these networks, idle wavelengths on a backbone trunk can be quickly allocated to other carriers or service providers through the implementation of optical switching systems. These systems allow an operator to treat the optical layer of its network much like it treats the ATM layer: as a pool of available bandwidth within a "cloud" to be quickly allocated in virtual circuits. In the case of optical networks, these virtual circuits are now optical circuits that are managed by optical switching systems using constraint-based routing algorithms. If vendors can develop optical edge equipment that can be agile enough with wavelengths, carriers might find it cost effective in certain instances to offer service providers or major corporate users the opportunity to purchase wavelength services not as a fixed lease or IRU, but as a flexible service. This would require a fully distributed metropolitan DWDM network in which a large percentage of the available interfaces on network equipment were installed and ready to be called into service by the network operator. Although this scenario is feasible in long-haul networks, it might not be in many metropolitan networks with limited DWDM deployments.

The class of optical-edge network gear that includes integrated DWDM functionality might allow some carriers to begin offering this service, although again it would require a widescale deployment of DWDM interfaces throughout a network. Today this comes at a cost of roughly $20,000 per DWDM interface, which is clearly cost prohibitive unless that interface is supporting a revenue-generating service from its initial implementation.

It is clear from all of this that multiple applications at the edge will drive a need for high bandwidth. If the needs for high bandwidth are not addressed, it will eventually lead to a total connectivity bottleneck. FSO can help service providers address this proactively. [2]

4

INTEGRATION OF
FSO IN OPTICAL
NETWORKS

FSO in Metropolitan Optical Networks

Now that you understand the overall MAN picture, you need to know how FSO fits into this overall hierarchy. The answer is simple. FSO is an optical technology that can address connectivity needs at any point in the network, be it core, access, or edge. FSO, with its capability to be Layer 1 and protocol transparent, is able to integrate with and interoperate with a variety of network elements and interfaces. This allows it to seamlessly be a part of the growing optical networking family.

Following are some of the common applications using free-space optics in MANs:

- Metropolitan network extensions: FSO can be deployed to extend an existing metropolitan ring or to connect new networks. These links generally do not reach the ultimate end user, but are more an application for the core of the network.

- Enterprise: The flexibility of FSO allows it to be deployed in many enterprise applications, such as LAN-to-LAN connectivity, storage area networking, intracampus connectivities, and so on.

- Last-mile connectivity: These are the links that reach the end user. They can be deployed in PTP, point to multipoint, or mesh connections. Fiber deployment in urban areas could cost $300,000–$700,000 given the costs involved in digging tunnels and getting right-of-way. By contrast, a short FSO link of 155 Mbps might cost only $10,000–$18,000 or as little as $166 per month (plus interest) on a 60-month amortization. This is a fairly monumental fact to grasp—the equivalent of three DS-3 lines for $166 per month! The present cost for three DS3s as leased lines from an ILEC could run as high as $10,000 or more per month!

- Fiber complement: FSO can also be deployed as a redundant link to back up fiber. Most operators who are deploying fiber for business applications connect two fibers to secure a reliable service plus backup in the event of outage. Instead of deploying two fiber links, operators could opt to deploy an FSO system as the redundant link.

- Access: FSO can also be deployed in access applications such as Gigabit Ethernet access. Service providers can use FSO to bypass local loop systems and to provide FSO-based high-capacity links to businesses.

- Backhaul: FSO can be used for backhaul such as LMDS or cellular backhaul, as well as for Gigabit Ethernet "off-net" to transport network backhaul.

- DWDM services: With the integration of WDM and FSO systems, independent players aim to build their own fiber rings, yet might own only part of the ring. Such a solution could save rental payment to ILECs, which are likely to take advantage of this situation.

Summary

POTS, SONET, wireless, first-generation optical networks, second-generation optical networks, and now free-space optics—this is quite a transition over a couple of decades. Although most of the other applications were new and disruptive changes in the telecommunications networks, free-space optics was not. Unknown to most people, free-space optics has been around for more than three decades, but interestingly enough, due to multiple market drivers, free-space optics has found a renewed value-added interest. It is fast becoming a value-added application for MANs that are enabling service providers to accelerate their deployment of optical networks, thus addressing the needs of their end users quickly and cost effectively.

With the evident growth in optical networks, it is clear that the connectivity bottleneck will continue to be shifting problems all across the optical networks. It is also clear that although innovation is key to such a growth, cost reduction is also a driving force. The all-optical network is focused on decreasing the cost per bit and making optical capacity available to the end users. Alas, some dreams are not easily realized, and the vision of the all-optical network finds itself in this dilemma of cost versus infrastructure.

To address and enable the acceleration of optical networks while addressing the need to be cost effective, free-space optics is presenting the users with an opportunity to do so. FSO is a perfect fit for the growing MANs fitting into multiple areas and not just last mile. Regardless of whether you use free-space optics in the core, access, or edge, one thing is clear: FSO addresses the connectivity bottleneck of today.

Sources

[1] These three paragraphs relating to storage area networks were taken from Chapter 3 of the report by Pioneer Consulting, LLC, "Optical Edge Networks: Market Opportunities for Integrated Optical Network Solutions in Metro Networks." August 2000.
`http://www.pioneerconsulting.com/report.php3?report=13`

[2] Much of the material presented in the VPN Services section through the Wavelength on Demand section was taken from Chapter 3 of the report by Pioneer Consulting, LLC, "Optical Edge Networks: Market Opportunities for Integrated Optical Network Solutions in Metro Networks." August 2000. `http://www.pioneerconsulting.com/report.php3?report=13`

4

INTEGRATION OF
FSO IN OPTICAL
NETWORKS

The FSO Market

IN THIS CHAPTER

FSO is not a new technology or invention. For several decades, FSO was used primarily in secure military communications. Now it has found commercial viability, driven by a number of factors:

- The cost effectiveness and minimal capital investment of FSO relative to other technologies
- The rise of high-bandwidth applications
- The rise of a global optical infrastructure
- Fiber shortages, with respect to both infrastructure and supply
- The ease of deploying and redeploying FSO solutions

This chapter takes a look at these and other factors driving the FSO market, and at the benefits of FSO relative to other technologies.

The Telecommunications Market

Before looking at FSO market particulars, this chapter will take a step back and look at related topics from the perspective of the larger carrier and enterprise network access market. This will allow you to better understand the strong market opportunity that FSO represents.

Optics Market

The optics market is growing and forecasted to hit $57.3 billion in sales of fiber-optic cabling, switches, routers, and related products and services by 2005. This market has been growing not only in the number of vendors and service providers, but also in technology adoption and deployment. Although some service providers are leading this race, others are slow to start because the infrastructure necessary to accelerate deployment is not yet available. Lead times for laying fiber are 14 months to 18 months, and many service providers have insufficient capital to buy fiber or make long-term commitments to invest in it. Meanwhile, the billions of dollars worth of fiber infrastructure that has been built is underutilized and essentially stranded.

With free-space optics, acceleration—not stranded—is the operative word. Free-space optics enhances and extends optical networks at substantially less cost than traditional cabling—and more importantly—at a fraction of the time. FSO allows high-bandwidth service to reach the end users much sooner than previously thought possible.

Free-space optics is not limited to the last mile. Instead, the technology is an enabler of optical networking. Free-space optics can play a role in the core, edge, or access, addressing segments such as last mile, Enterprise/LAN, or metropolitan network extension. FSO is flexible and fast (speeds of 1 Mbps to 2.5 Gbps). Imagine all the places where barriers (variables of cost, time, and physical obstacles) prevent a service provider from offering optical services. Free-space optics offers a cost-effective, quick, and available infrastructure that is not only easily deployed (within hours), redeployed, and easy to manage, but can also offer a multitude of options, such as distance, speed, topology and installation flexibility. And free-space optics can do all of this with carrier-class features of fiber level throughput, high reliability, availability, and multiprotocol capability.

The broadband network has a bottleneck. It exists primarily in the last mile and metropolitan areas where, due to high cost per throughput unit, it has not been profitable for broadband carriers to provide services. At present, a mixture of copper, fiber, and even some RF fixed wireless solutions are in place, but in the future addressing this bottleneck, unlike others, will involve true optical networking (as opposed to a combination of electrical and optical networking). FSO allows service providers to accelerate their deployment of metropolitan optical networks as well as extend the reach of such optical capacity to anyone who needs it. What's more, FSO delivers this optically without digging trenches or buying expensive spectrum.

Demand for bandwidth has been increasing exponentially for the past few years. Service providers have been struggling to keep up with such demand. Although a tremendous effort is under way to upgrade the core, the same effort has not been made for end users. In fact, end users, in many cases, feel abandoned. With all the bandwidth that will be available in the metropolitan core, service providers must find a way to reach the end users. Service providers must extend the reach of metropolitan optical networks, and FSO offers service providers the opportunity to do so more economically and on a shorter timeline than any other technology with comparable carrier-class qualities.

Broadband Market

Broadband communications have been a large driver in the growth of high-bandwidth applications. Broadband is a direct result of growth in the Internet, intranets, and the increased proliferation of voice, video, and data applications, and FSO is a direct result of the need to accelerate the reach of high-speed networks to end users.

Internet users have grown from 153 million worldwide in 1998 to 320 million in 2000 [1]. Data traffic on the Internet backbone has roughly doubled every year in that time and continues to do so. Despite some hiccups in the market, PC shipments are expected to continue to grow at 10% to 15% annually, reaching 257 million units in 2005. Meanwhile, e-Commerce is already estimated at well over $1 billion, and shipments of low-cost Internet appliances are poised to explode.

The success of the online world is clear, but expansion in the number of services and quality of experience that the Internet provides is just beginning. The key is to provide this next level of high-bandwidth access to end users by extending the reach of the core metropolitan optical networks.

Service Provider Needs

With a growth in end user needs for immediate high bandwidth and the inability for network deployments to catch up to meet those needs, service providers find themselves stranded. Service providers are stranded both in terms of invested capital and invested resources with no revenues being generated by the invested capital. Further aggravating this challenge is the increasing number of service provider competitors. Service providers must alleviate capital concerns and identify tools that allow them not to just stay ahead of the competition, but to remain in business. FSO addresses many service provider needs, as discussed in more detail in Chapter 8, "Service Provider Issues."

Trends in Bandwidth Demand and Availability

It is clear that bandwidth usage is experiencing unprecedented growth, and bandwidth demand is not likely to slow down in the near future. A number of factors are influencing this surge in bandwidth demand, and the following are most compelling:

- Internet growth: In 1999, Internet traffic alone accounted for some 350,000 TB per month of traffic, and that traffic is expected to increase 40-fold, up to 796 million users by 2005. Broadband service subscribers are expected to grow to 123 million by 2005. All this adds up to more bandwidth being pushed to the edge.

- Multimedia edge: With an increasing number of users of multimedia applications in businesses, the need for high bandwidth at the edge has risen to unprecedented levels, leaving many service providers stranded for capacity. Short of leasing and releasing lines at a loss, many service providers have no alternative but to refuse service.

- Changing traffic patterns and protocol standards: Metropolitan networks are characterized by multiple traffic types. Where voice was once the dominant traffic type, convergence is well underway, and voice, video, and data increasingly share the same infrastructure. Moreover, the networks must support a mixture of protocols ranging from Ethernet, SONET, IP, ESCON, FICON, and so on.

- International growth: Most countries are experiencing tremendous growth in bandwidth needs, due to the growing number of Internet-based applications.

- Wireless world: The wireless world is driving bandwidth demand as more and more applications become accessible through the wireless infrastructure.

Characteristics of the FSO Market

FSO is in a position to address many of the bandwidth trends just discussed in a manner that is competitive with or superior to fiber optics, wireless, and other technologies. This section discusses some of the factors that will enable the growth of FSO and address the total market potential for FSO.

FSO Challenges and Benefits

To have a complete understanding of the market for FSO, it is important to understand the challenges as well as benefits of the FSO technology.

Primary FSO challenges include the following:

- People's perception of it as a niche application
- People's lack of awareness about FSO
- Existence of competing alternatives such as copper and fiber lines as well as RF fixed wireless
- Atmospheric conditions such as fog and rain
- Line of sight limitations
- Distance limitations
- Reliability of FSO versus competing alternatives
- Slow adoption process

On the other hand, FSO offers some distinct advantages:

- Cost effectiveness
- Quick to deploy and redeploy
- Environmentally safe, allowing for balance of ecosystems, technology, businesses, and communities
- Does not require "reinventing the wheel" for adoption into existing networks, and it is a transparent technology, making it easy to integrate
- A "build as they come" model, entailing no sunk costs
- Scalable with respect to bandwidth, from 1 Mbps to 2.5 Gbps
- Highly secure (was first developed for secure military communications)
- Minimizes capital requirements for a buildout
- Reliable equipment, with MTBFs of up to 23 years quoted by some manufacturers
- Many models function like a bridge and are protocol independent

Market Size and Growth Predictions

Currently, the FSO market is in a growth stage. For 2001, the market is expected to reach a total of $250 million compared to $118 million in 2000[2]. The global market for free-space optics-based applications is anticipated to be 2–4 billion by 2005. Globally, this forecast includes the following regions: North America, Europe, Asia Pacific (China, India, Korea,

5

THE FSO
MARKET

Japan, Taiwan, Indonesia), and Latin America. The primary driving applications are metropolitan network extension, access, or last mile and enterprise connectivity.

Additional factors that will influence the growth of the FSO sector are growth in the metropolitan optical networks and technology innovation leading to more applications that would require cost-effective high-speed connectivity. Such applications include IP telephony, data warehousing, off-site data backup, Application Service Providers (ASPs), Virtual Private Networking (VPN), and increased e-Commerce.

Market Segments

FSO is a part of the metropolitan optical networking market of the optical industry. The MON market is further categorized by metropolitan core, metropolitan access, and metropolitan edge. FSO, with its capability to transmit bandwidth up to OC192/10 Gbps, qualifies as a short-distance metropolitan application.

To better understand how FSO will provide bandwidth solutions, consider it by market segments as follows:

- By application
- By customer type
- By region

By Application

Segmentation by application helps to further define how FSO can be applied to each application in terms of needs and requirements. These applications fall under seven main categories. They are:

- Enterprise/LAN: Interconnection of corporate /campus networks.
- Redundant link and disaster recovery applications: This is a specialized application used to provide backup systems for mission-critical connectivity. Categorizing it under Enterprise/LAN would be unfair, given the growing size of such applications and also the variety of protocols used (Ethernet, FDDI, ESCON, FICON, and Fiber Channel).
- Storage area networking (SAN): This emerging market finds itself particularly bandwidth-constrained. FSO offers the same value to storage area networks that it does to other networks: the capability to have networks up and generating revenue cost effectively and reliably.
- Last mile access: A significant bottleneck exists at the last mile of most service providers' networks. The challenge is to connect paying customers to their vastly under-utilized worldwide fiber backbone. This access solution deals with point-to-point connectivity for distances ranging from 50 m up to a couple of miles.

- Metropolitan network extension applications: These applications enable existing core networks to achieve greater capacity and reach, as well as provide the capability to fill gaps in the network. These include extending the network through a mixture of technologies and standards, such as SONET, GigE, ATM, and IP. The applications addressed under this category include SONET ring closures, spurs from existing rings for access connectivity, F1/F2 relief, and wireless backhaul, among others.

- DWDM networking applications: These are "pure optical" solutions where the network communicates with the edge optically without O-E-O (optical-electrical-optical) conversion.

- Backhaul and augmentation of next-generation wireless networks—3G and 4G: These represent a segment of the market that is capital intensive and has been slow to deploy primarily due to ROI barriers. FSO offers a vehicle that would allow service providers to accelerate this deployment and achieve their revenue targets. The primary application of FSO will be in backhaul and augmentation of 3G/4G networks.

By Customer Type

Segmentation by customer type provides a means of understanding the varying needs of each customer. An FSO vendor's target customers in the metropolitan/access carrier market include an array of different service providers:

- Enterprise customers remain a large segment for LAN extension, redundant link, and disaster recovery.

- Next-generation Inter-Exchange Carriers (IXCs): This category of customers are long-haul service providers who are making forays into metropolitan networks.

- ILECs: This group represents the customer type with the largest networks and largest capital budgets. They are slow to make decisions, but have large-scale deployments. They are also known as RBOCs (Regional Bell Operating Companies).

- Broadband CLECs (Competitive Local Exchange Carriers): These companies are a result of the Telecommunications Deregulation Act of 1994. Their mission is to provide competing services to ILECs; however, due to increased pressures, most of the CLECs have either gone out of business or have been absorbed by bigger players.

- Onsite service providers and commercial real estate owners and managers: As the name suggests, this category of customers are owners who build "broadband ready" buildings, providing their tenants high-speed connectivity.

- New metropolitan carriers: This group represents customers who are aggressively building fiber-based networks, such as MFN and Looking Glass networks.

- Utility companies moving toward a telecommunications company model: This represents a growing base of customers. With access to millions of customers, utilities have been slowly venturing into the telecommunication market.
- Wireless service providers: This group delivers Internet access as well as high-bandwidth "last mile" solutions.

By Region

Segmentation by region will help you understand where to deploy resources and what parts to focus on for maximum return on your deployments. Segmentation by region is driven by the lifestyles of customers, governmental and regulatory issues, the status of infrastructure, and the availability of alternatives. It also is instructive for understanding the reasons that international deployments of FSO will be substantial.

FSO growth in North America will be driven in part by the pressure faced by service providers to do the following:

- Acquire more customers
- Generate revenue quickly
- Offer services cost effectively

North America is anticipated to be a relatively slow-growth region primarily due to longer sales cycles and the availability of other alternatives. These combined factors contribute to slow adoption of a new technology on a large scale.

Due to a more aggressive approach to new technology adoption as well as lack of alternatives, major growth in FSO is anticipated to primarily be in Asia (China, India, and Southeast Asia), Latin America, and Europe. This growth will be driven by the following factors:

- Lack of infrastructure
- City authorities who are less likely to allow service providers to dig up congested streets
- Need to have services up quickly
- Less stringent availability requirements
- Dense and bigger metropolitan areas and more buildings, resulting in shorter distance spans
- Unavailability of bandwidth

Figure 5.1 depicts projected U.S. versus international FSO market growth.

FSO Drivers

As Figure 5.2 shows, a number of key interrelated drivers are at play in the FSO market. This section discusses each of them briefly.

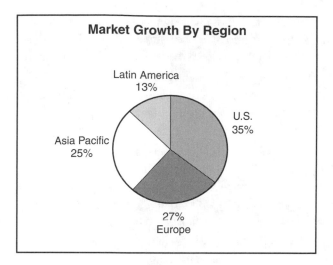

FIGURE 5.1

Market growth by global region.

FIGURE 5.2

A number of drivers are pushing the growth of FSO.

Market Drivers

Market drivers are those factors that are driving growth of FSO based on the overall influence of market conditions:

- Increasing number of Internet users/subscribers: As noted earlier, the Internet is causing a demand for high bandwidth at the edge of the network. The number of Internet users is expected to grow to about 796 million by 2005.

5

THE FSO MARKET

- Increasing e-Commerce activities: With a growing number of businesses involved in B-to-B activities, e-Commerce is fast becoming a user of high bandwidth. To meet this huge increase in bandwidth, service providers must offer high-bandwidth access at the edge of the network.

- High-capacity desktops: With increased deployment of multimedia applications and the continuing exponential increase in processor speeds, the desktop is now an enabler of high-bandwidth applications.

- Deployment of metropolitan optical networks: Service providers are investing millions of dollars in deploying DWDM-based metropolitan optical networks. Upgrading metropolitan optical networks is a direct result of the increase in bandwidth usage at the edge. Service providers are now faced with mounting pressures to speed up this deployment, while at the same time generate revenues. FSO meets both of these needs.

- MMDS/LMDS: Given the high cost of acquiring spectrum, the time needed to build such networks, along with the fact that such systems need to be linked to each other and to the PSTN, FSO offers an alternative connectivity path with specific benefits.

- Commercial buildings without high-bandwidth access: Given the fact that early networks were developed to meet the needs of the core, and the majority of commercial buildings were constructed in the early 1980s, a large majority of those buildings do not have high-speed access. The statistic frequently quoted is that 95% of U.S. businesses are not serviced by fiber, and 75% of those are within one mile of fiber. FSO provides a quick, cost-effective, and reliable means to address the needs of those users.

- Deployment of 3G and 4G (advanced digital wireless phone services): Spectrum scarcity coupled with bandwidth appetites in metropolitan networks are forcing wireless operators to look at new methods to connect cells. FSO offers a viable option.

Economic Drivers

Economic drivers of FSO growth impact the profitability of a company. The primary economic drivers are as follows:

- Reduce costs: With costs that are considerably lower than traditional equipment, FSO offers service providers the opportunity to reduce their costs immediately. The simple reason that FSO costs less than competing connectivity solutions is that it is basically fiber-optic connectivity without the fiber. This effects a substantial cost advantage: no digging or trenching, lower installation costs, and the equipment is rather inexpensive.

- Faster service activation: With installation times as low as 4 hours, service providers can turn up services quickly, thus generating revenue quickly.

- Ultrascalability of bandwidth allowing for lower inventory costs: With a scaleable technology generally covering 1 Mbps to 2.5 Gbps, FSO offers a broad range of speeds that is scaleable within a matter of hours to meet customer needs.

- Multiple applications/services support: Using FSO products that are Layer 1, service providers can offer multiple services (GigE, Fiber Channel, ESCON, and so on) from the same platform due to the inherent transparent core of the product. The layer approach makes FSO just like fiber.

- Quicker time to market: Easy installation and quick deployment of FSO products allows service providers to turn up new services virtually overnight. This can be the difference between success and failure in the current highly competitive marketplace.

Service Drivers

Service drivers are those factors that contribute to FSO's flexibility, ease of integration, and user-friendliness. Service drivers include the following:

- Increasing demand for high-speed access interfaces: The interface flexibility that FSO provides, such as OC-48, GigE, ESCON, FICON, and so on means that FSO will be able to meet this increasing diverse demand for high-bandwidth applications quickly and cost effectively.

- Need to eliminate the metropolitan gap: FSO helps avoid situations in which the network edge and core work against each other. With DWDM and wavelengths operating at 1,550 nm, service providers will be able to integrate metropolitan optical networks with FSO networks.

- Network simplicity: Fewer network elements (since the linkhead is a single unit) mean fewer points of failure and fewer elements to manage.

- Need for real-time provisioning: With changing customer needs and unpredictable traffic patterns, service providers need the ability to provision on demand. The scalable, Layer 1 FSO approach, being transparent and able to provide bandwidth up to 2.5 Gbps, enables service providers to accomplish that.

Business Drivers

These days, business case development has zero tolerance for any "build-it-and-they-will-come" mentality. Infrastructure must have cash flow in Internet time (immediately). FSO really packs some punch here! Business drivers include the following:

- Accelerating metropolitan optical networks: FSO can help service providers accelerate and extend their metropolitan optical networks to the end user.

- Customer retention: In this dynamic and abrupt competitive environment, FSO offers service providers a tool that helps them stay competitive and ahead of competing technologies.

- Variable SLAs: To globally compete with other services and services providers, variable SLAs are valuable for many service providers. Variable SLAs offer varying levels of customer satisfaction. With FSO, service providers can choose to offer SLAs for each deployment in each geographic location.

- Global lack of infrastructure: With growing economies around the world, many countries use equal or more bandwidth than North America. This shift has created a definite need for a high-bandwidth platform. FSO offers a unique opportunity with its license-free capacity, along with its low-cost solution to meet those emerging needs.

Environmental Drivers

Environmentally, FSO is gentler than all competitive technologies. Environmental drivers include the following:

- Pollution avoidance: Digging streets to lay fiber causes traffic jams in major cities, contributing to pollution. With FSO, you do not have to dig up streets and add to the pollution.

- Preservation of historic landmarks: In some older communities, digging up streets might mean destroying some historic landmarks that might never be recovered.

- Ecosystem: Laying fiber also translates into sometimes cutting trees. With FSO, one does not have to disturb or destroy the ecosystem, thus preserving the natural balance.

Technology Drivers

Technology trends also are driving FSO:

- The Internet has fast become an integral part of today's business and consumer sector. The proliferation of the Internet in those lifestyles will further drive the need for high-bandwidth connectivity.

- The optical shift in the telecommunications networks—especially in the metropolitan core—has further made it essential to enable optical technology at the edge of the network. FSO clearly enables such connectivity.

Adoption and Implementation

Despite high unmet demand for bandwidth at the edge of the carrier networks, FSO will undergo a period of gradual deployment as it faces the challenge of gaining widespread market acceptance. This challenge is not unique to FSO, but it is characteristic of most new technologies. It is valuable to understand the FSO adoption cycle.

The direction of FSO migration is from the enterprise market (global) to carriers and CLECs (globally), to RBOCs, and finally to the residential markets. The leaders in the FSO market are currently at the carrier-adoption phase, having proven successfully at the Enterprise level with customers and deployments. That clearly sets apart the leaders from all others. To move ahead and prove the product successfully to carriers, the FSO vendors need to incorporate certain features in their product line. Chapter 8 provides a detailed discussion of issues that FSO must address to become a successful "carrier class" technology.

Armed with the appropriate features, FSO will become a part of the carrier toolbox, thus finding its way into many carrier applications, such as last mile, metropolitan network extension, network redundancy, or an optical gap filler wherever high capacity is needed quickly and cost effectively. After FSO migrates into the core network, it clearly enters the FSO shift zone.

The FSO *shift zone* (see Figure 5.3) is the point where FSO becomes mainstream. Standards begin to become important and RBOCs feel pressure to start deploying FSO into their networks while developing a global support infrastructure ensuring interoperability and integration with their systems. Innovation will drive the pricing down and further drive the application closer to the ultimate end user. If forecasts are correct, that end user will be the ultimate driver of high-bandwidth applications.

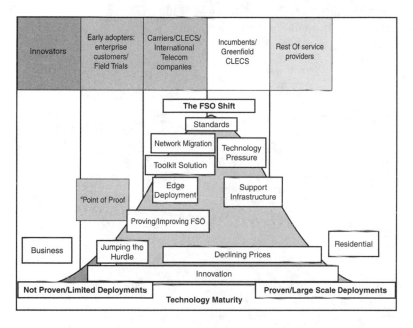

FIGURE 5.3

FSO "shifts" to incumbents such as the regional Bell operating companies after enough competitive and international carriers adopt the technology and it begins to appear in the network core. Source: LightPointe Analysis.

The Business Case for FSO

So you can get a feel for the range of business cases for FSO, this section analyzes three applications: Gigabit Ethernet for backhaul, DS-3 services, and Sonet Ring closure. These analyses are presented in the form of case studies.

5

THE FSO
MARKET

> **NOTE**
>
> For all the cases in this section, FSO equipment costs and fiber-related installation costs are based on industry averages rather than specific vendors' prices.

Case 1: Gigabit Ethernet—Access and Backhaul

A competitive carrier signed an agreement with a large property management firm to provide all-optical, 100 Mbps Internet access capability to several buildings located in an office park. The carrier is building its network by leasing regional dark fiber rings and long-haul capacity from a wholesale fiber provider. The carrier is evaluating FSO to connect "off-net" buildings and also for backhaul to the transport network. Figure 5.4 shows the customer site in relation to the Regional Ring and the Metro Transport Ring.

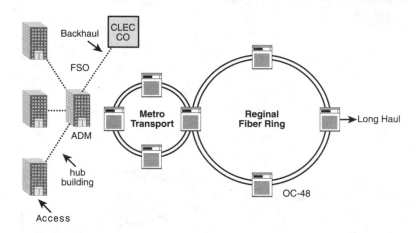

FIGURE 5.4
Example of LAN extension and backhaul FSO deployment.

Pertinent details of the customer's existing network architecture and requirements include the following:

- No fiber currently is deployed to target customer buildings.
- Competitive carrier facilities are located 1 km from office parks.
- Native transport of Gigabit Ethernet is required.
- Excess capacity is desired for future advanced services and new customers.

The service provider's business requirements are as follows:

- Provide 100 Mbps Internet access to building customers at $1,000 per month
- Penetration rate: three customers per building
- Turn up new services next quarter (three months)
- Generate a minimum 30% Internal Rate of Return (IRR)

For the building access portion of the network, the service provider is evaluating new fiber deployment and FSO 1.25 Gbps/1 km. For the backhaul to the transport network, the options are new fiber deployment, dark fiber lease, or FSO (again, 1.25 Gbps/1 km). The access and backhaul options will be considered in turn.

Building Access

The competitive carrier does not have fiber deployed to the office park. In addition, no fiber (dark or ILEC-owned) is available for lease. This leaves only two options: (1) new fiber deployment or (2) free-space optics.

Table 5.1 compares the deployment costs of fiber and FSO for service to three buildings located an average of 500 m from the carrier's "hub" building.

TABLE 5.1 Fiber Versus FSO Deployment Costs for Building Access

Fiber	
Total distance from POP/hub building to off-net building	500 meters (.3 miles)
Total number of feet (5280/3) * Sqrt2)	2,489
Percent trunk fiber	70%
Total feet: trunk fiber	1,742
Cost per foot	$100
Total cost: trunk fiber	$174,230
Percent feeder fiber to three buildings	30%
Total feet feeder fiber	747
Cost per foot	$100
Total cost feeder fiber to three buildings	$224,010
Total fiber deployment costs: Building access	**$398,240**

TABLE 5.1   Continued

FSO	
FSO equipment cost per building	$18,000
Installation per building	$5,000
Total FSO cost: Building access	**$69,000**

Table 5.2 shows the internal rate of return analysis for fiber versus FSO building access. The assumptions for this analysis are as follows:

- In Year 1, the carrier serves three buildings with three customers per building.
- The carrier charges each customer $1,000 per month for 100 Mbps Internet service.
- The carrier anticipates a 15% annual revenue increase for advanced services and new customer acquisition.

TABLE 5.2   Internal Rate of Return (IRR%) for Building Access: Case Study 1

	Fiber	FSO
Capital Investment	($398,240)	($69,000)
Cash Flow Year 1	$108,000	$108,000
Cash Flow Year 2	$124,200	$124,200
Cash Flow Year 3	$142,830	$142,830
Cash Flow Year 4	$164,255	$164,255
Cash Flow Year 5	$188,893	$188,893
IRR%	22%	196%

Backhaul to Transport Network

For backhaul to the central office (CO), the carrier will transport Gigabit Ethernet natively, due to the high electronics/equipment costs of transmitting Ethernet over SONET. The carrier has a CO within 1 kilometer from the hub building and is planning to backhaul traffic to the CO based on a point-to-point 1 km connection using either (1) carrier-deployed fiber, (2) leased dark fiber, or (3) free-space optics (1.25 Gbps/1 km). Table 5.3 compares the capital investment required for each technology.

TABLE 5.3 Comparison of Backhaul Costs

Fiber	
Cost per foot (U.S. Central business district)	$100
Feet per mile	5,280
Miles in ring	1
Total fiber deployment cost: 1 mile, point-to-point connection	$528,000
Dark Fiber Lease	
Cost of dark fiber lease per mile per month (U.S. Central business district)	$1,200
Number of fibers	2
Ring distance in miles	1
Length of dark fiber lease contract (IRU) in months	60
Total dark fiber lease cost: 1 mile point-to-point connection	$144,000
FSO	
Cost of equipment (1 km/1.25 Gbps	$52,000
Installation and setup	$7,000
Total FSO cost	$59,000

Payback Period

Figure 5.5 compares the payback period of a network based on an FSO solution, for both backhaul and access, versus an all-fiber network, using leased dark fiber for backhaul and newly deployed fiber for access.

Case 2: DS3 Services

An incumbent local exchange carrier receives more than 4,000 requests for DS3 service in a 12-month period but is able to fulfill only 60% of these service requests. A new carrier entering the local market plans to capture a significant portion of this business using FSO.

	Capital Investment	Cash Flow, Dollars					Payback Period (moderate)
		C1	C2	C3	C4	C5	
Fiber							
Access: Fiber Deployment	$298,680						
BackHaul: Dark Fiber Lease	$144,000						
TOTAL	$442,680	$108,000	$124,200	$142,830	$164,255	$188,893	3.5 years
Free Space Optics							
Access	$59,000						
Backhaul	$52,000						
TOTAL	$111,000	$108,000	$124,200	$142,830	$164,255	$188,893	13 months

Figure 5.5

Payback analysis for Case 1.

The competitive carrier deployed a 155 Mbps FSO connection on the rooftop of a line of sight customer building. For each DS3 customer captured, the carrier will deploy an OC-3 multiplexer in the building basement to capture additional DS3 or T1 business. Figure 5.6 shows the payback schedule for the FSO investment.

DS3 Deployment	Capital Investment	Revenue M1	Revenue M2	Revenue M3	Revenue M4	Revenue M5	Payback Period
FSO Costs							
Equipment (155mbps/600meters)	($18,000)						
Building Infrastructure/Set-up	($6,000)						
OC-3 Mux	($20,000)						
TOTAL FSO Costs	($44,000)						
Monthly Revenue per DS-3		$4,000	$4,000	$4,000	$4,000	$4,000	
Number of Customers		1	1	1	1	1	
TOTAL Revenue		$4,000	$4,000	$4,000	$4,000	$4,000	11 months

Assumptions
DS3: $4000 per month
33% penetration rate per building

Figure 5.6

Payback analysis for Case 2.

Case 3: SONET Ring Closure

A carrier has an OC-48 SONET ring with two OC-12 fiber spurs to connecting buildings (see Figure 5.7). To offer building customers "on-ring" service-level agreements (SLAs), the carrier is planning to offer physically diverse SONET protection through a closed ring architecture. A larger, noncustomer building obstructs the two customer buildings; therefore, the service provider must deploy the ring *around* this building. The carrier is evaluating FSO equipment (622 Mbps/1 km) and fiber for ring completion.

The carrier deployed a 622 Mbps/1 km FSO connection to close the fiber SONET ring and provide physically diverse protection. This enabled the carrier to offer higher service-level agreements and generate new revenue streams.

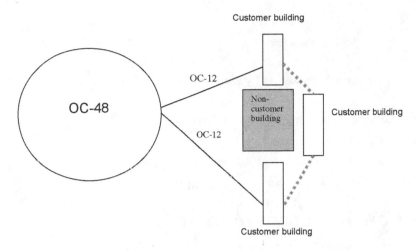

FIGURE 5.7
Fiber or FSO will be used to complete the SONET protection ring.

A carrier providing physically diverse protection to a building charges, on average, an additional $3 per square foot on an annual basis versus a building without protected facilities. For example, a carrier servicing an MTU with 35,000 square feet can generate $105,000 of additional revenue by providing physically diverse protection with free-space optics and realize a payback period of approximately 11 months, as detailed in Figure 5.8.

SONET Ring Extension	Capital Investment	Revenue Y1	Revenue Y2	Revenue Y3	Revenue Y4	Revenue Y5	Payback Period
FSO COSTS							
Equipment							
(622 Mbps/1km)	($45,000)						
Number of Links (bypass obstruction)	2						
TOTAL Equipment Cost	(90,000)						
Building Infrastructure/Set-up	($6,000)						
TOTAL FSO Cost	**($96,000)**						
REVENUE							
Size of Building (Sq ft.)		35,000	35,000	35,000	35,000	35,000	
Additional Revenue per sq ft. (annual)		$3	$3	$3	$3	$3	
Annual Revenue		$105,000	$105,000	$105,000	$105,000	$105,000	11 months

FIGURE 5.8
Payback analysis for Case 3.

Conclusions from Business Cases

It is clear from the preceding application analysis that FSO offers service providers a business case that is a credible alternative to laying fiber. By deploying FSO, service providers can potentially capture these benefits:

- Increase profit level on existing capital. For example, providers can extend existing fiber or LMDS network without additional equipment/training/licensing costs.

- Increase return on new capital investment.
- Offer High Margin Services, such as 2.5 Gbps, which enables more versatile service offerings.
- Grow their customer base quickly by acquiring new customers or leveraging current on-net buildings.
- Generate revenue by bringing off-net buildings on-net quickly.
- Reduce cost of capital.
- Conserve capital by taking the build-as-they-come approach.
- Have zero sunk costs because FSO is a redeployable platform.

International Telecom Market

Because market acceptance of free space optics is accelerating rapidly outside North America, it is important to understand international telecom markets. The international telecommunications markets that are experiencing the highest growth are Europe, Latin America, and Asia Pacific (primarily India, China, Malaysia, and Indonesia).

Telecommunications-related spending in Canada, Latin America, Europe (east and west), and Asia Pacific totaled an estimated $1.2 trillion in 2000. This was 17.5% higher than 1999. The telecommunications services included in this growth are transport services, equipment, support services, and wireless services.

Such increased spending in these regions represents tremendous growth potential in various segments of the telecommunications market. It is estimated that the transport services will continue to grow at a steady rate because that infrastructure needs to be built to enable connectivity in these growing economies (especially Asia Pacific). With no fiber infrastructure and not so much abundance of copper infrastructure, wireless services will continue to grow. To enable these transport and wireless networks, both the equipment and support services will generate considerable revenue. All in all, the telecommunications sector outside North America will experience consistent growth.

The same drivers mentioned earlier in this chapter apply to international service providers. They will have to find ways to enable high-speed connectivity not only quickly, but also cost effectively. With high-speed connectivity needs on the rise and lack of infrastructure (both copper and fiber), it is imperative for service providers to look for alternative technologies. This book has discussed alternative technologies both from a technology and economic aspect in prior chapters, and it seems evident that free-space optics really has the potential to address the needs of these regions.

The international regions mentioned previously are highlighted by the following key characteristics:

- Older buildings: The majority of the buildings in these cities are more than 30 years old. It is safe to assume that most of them are not connected to any type of fiber infrastructure, and that very few of them have copper connectivity. FSO can play an important role in bringing these buildings on-net and providing high-speed connectivity to these enterprises.

- Dense urban areas: Most developed metropolitan areas in these regions are dense. This means that buildings are closer. After you bring one building on-net, it is easy to extend that connectivity to the next building. Furthermore it makes building-to-building connectivity easier. In such conditions, FSO offers a unique opportunity to carriers and enterprises to have optical connectivity.

- Moratoriums: Most governments in these regions have imposed moratoriums on digging. This is driven primarily by increased traffic and underdeveloped transport infrastructure (such as poor roads). With these conditions, service providers have to look for an alternative that can enable connectivity without trenching or digging. FSO fits the bill perfectly.

- Lack of infrastructure: Due to lack of physical infrastucture (copper or fiber), it is almost impossible for service providers to provide high-speed connectivity to its customers. This leaves providers no choice but to use free-space optics as the means to provide this connectivity.

- Need for high-speed connectivity: With the growing economy and increased number of businesses, high-speed connectivity is fast becoming a necessity instead of a privilege. More companies are looking for means to enable high-speed connectivity, and needless to say, FSO offers the solution they need.

Clearly, international markets will experience significant growth in high-speed services. Not many cost-effective alternatives are available to address these needs, and none of them offer the benefits that free-space optics offers.

Summary

Currently, FSO is regarded as a niche technology, but it is only a matter of time before it moves from niche to mainstream. In the short term, FSO will continue to address the immediate needs of both enterprise and carrier customers across many segments of the communications market. But as more systems are deployed and carriers become comfortable with and convinced about the reliability of FSO, the market for FSO will increase significantly. FSO will then move from a niche to a core technology. From a global perspective market research institutions are suggesting that the market outside the United States will be much larger than the domestic market [2].

Sources

[1] Allied Business Services (www.alliedworld.com)

[2] The Strategis Group

Installation of Free-Space Optical Systems

IN THIS CHAPTER

Like most systems, correct installation of an FSO system is extremely important for its functioning and long-term stability. If installation is done properly, FSO systems will function without fail for a long time. Like most technologies, FSO has its own set of operational requirements, some of which are unique to laser-based equipment.

How do you properly plan and deploy FSO links? Planning involves assessment of your needs from both the customer's perspective and the network perspective. Evaluating your network requirements is critical to determining which vendor's products to use, and evaluating your customers' requirements enables you to determine what type of FSO system to buy (what rates, what interfaces, and so on).

Deployment involves assessment of the environment, including weather patterns, link distances, line of sight, and so on. Proper analysis of both customer and network needs is extremely important to ensure successful installation of your FSO network. These seemingly straightforward considerations can easily be dealt with. Poor planning will lead to impaired link performance and availability over the long term.

Installation of FSO systems incorporates multiple steps, but after the planning is completed, actual installation takes only 2–4 hours. The main steps involved in any FSO installation include the following:

- Obtaining the site survey
- Mounting the equipment
- Installing the infrastructure (cabling, electrical)
- Aligning the systems
- Verifying the link

Obtaining the Site Survey

Site surveys are one of the most important steps of installing an FSO system. Site surveys can be performed either by a trained technician of the service provider or by trained personnel of the FSO manufacturer. Site surveys involves gathering of information that is important to know prior to any installation. FSO manufacturers will provide the necessary site survey documents.

A site survey document typically is a questionnaire designed to obtain answers to the most important questions before any installation or shipment of equipment by the vendor. Questions include:

- Is there line of sight?
- What kind of power supply is requested?
- What is the application like?
- What are the addresses?

Installation of Free-Space Optical Systems

CHAPTER 6

125

6

INSTALLATION OF
FREE-SPACE
OPTICAL SYSTEMS

Site survey is a series of steps that, when followed, will yield information that will lead to a successful deployment.

Most FSO vendors provide site survey training with the initial sale of the system. The goal is to make the customer self-sufficient so that he can perform his own site surveys. Surveys are not a complex art, but rather a detailed one. Some FSO vendors charge for site surveys either as a training seminar or a per-site survey. These fees typically range between $300–$2,000 depending on the scope of the project, such as number of sites.

The first step toward conducting a site survey is a visit to the site of the deployments. The following sections outline the typical suggested and required information to be obtained during the site survey.

General Configuration of the Sites

The first on-site order of business is to determine the general configuration of the sites. It is helpful to initially develop a sketch of the buildings, documenting their positions and aspects, and later documenting the distance between the two points. In most cases, the customer will have a network schematic of the proposed link or links. Put as much detail as you can into such schematics. In most cases, vendors request actual pictures of the deployment sites.

General Information

Next, verify that you have the exact address of each of the sites. This is important to determine the link distances as well as the line of sight from each location. After you verify the location, distance, and line of sight, the next step is to determine the proposed mount location on the buildings. Determining the mount location and type of mount are among the most important steps of the site survey.

When considering deployment locations, you have two choices: rooftops or windows (see Figures 6.1 and 6.2). When deploying behind windows, check the manufacturer's rated link distance reduction for tinted and untinted glass and also for minimum and maximum angle to glass. Deployment behind a window can be a favorable choice because it can reduce some outside weather impacts, such as snow buildup. It also can reduce cost due to the closer proximity of line power and the network interface, the elimination of the need for lightning protection, and the avoidance of paying for roof rights. Overall, you can reduce costs by deploying behind windows.

Obstacles to an inside mounting include heavy window tinting or potential for window frost, danger of interference with the equipment by personnel (including window washers), and inadequate equipment elevation with respect to line of sight or atmospheric scintillation considerations.

FIGURE 6.1
A rooftop installation.

Line of Sight

Free-space optics is a line of sight technology; in other words, if you can see the end you are connecting to, you can link it. Sounds simple, but of course it's a little more complicated than that. Multiple considerations go into determining line of sight, some of which are easy to point out, whereas others have to do with applying considerable foresight to your environment. One of the elementary steps used in the process of line of sight determination is taking pictures of the proposed link locations from one end to the other and then analyzing them (see Figure 6.3). Such pictures will help you consider both existing and potential obstructions such as tree growth, new construction, intermediate rooftops, chimneys and smokestacks, and flagpoles that might have a flag flying tomorrow. Some of these obstructions are easy to overlook unless you take the time to do a careful analysis.

Due to the dynamic nature of the environment, determination of line of sight is not an exact science, but more of an exercise in analysis and forethought.

Installation of Free-Space Optical Systems

CHAPTER 6

127

6

INSTALLATION OF
FREE-SPACE
OPTICAL SYSTEMS

FIGURE 6.2

A behind the-window installation.

FIGURE 6.3

Line of sight visual.

Link Distances

One of the steps of the site survey is determining the link distances. Link distance is measured as the distance between the two link heads. It is important to know this distance because the equipment can be distance limited. Distance limitation is more acute in adverse weather conditions such as dense fog, which could shift the availability figures of that particular link. Another issue that is affected by distance is power. Overpowering at short distances can potentially cause signal saturation at the receive end, thus rendering the link nonfunctional. On the other hand, underpowering can lead to link failure due to decreased link margins at longer distances.

The two methods for measuring link range are a measuring wheel or a GPS device. The latter is preferred because it is the only means of measuring straight-line distance if structures or obstacles are in the path that would have to be walked around using a measuring wheel. Another method to measure short-link distances is to use a laser viewfinder.

Mounting Considerations

After the location for mounting of the FSO link head has been selected, attach the mounting base plate to a solid platform. Whether the location is on a rooftop, on the side of the wall, or behind windows, it is important to have a stable solid platform. It is important that the mount point be stable because any fluctuation in the mount can cause link misalignment. Concrete and masonry structures are better than steel, steel is better than wood, wood is better than external Styrofoam sheathing, corners are better than mid-sidewall, mid-sidewall is better than horizontal roofing.

Each FSO system comes with a universal mount that is used for standard mounting purposes. Mounts are designed to prevent corrosion (aluminum, stainless steel, galvanized or powder-coated steel), and must be strong and provide for rigid building attachment. Two typical rooftop mounts exist:

- Penetrating mounts: These mounts are bolted to the roof.
- Nonpenetrating mounts: These are mounts that are held in place with the aid of heavy weights, such as sandbags or water containers.

Tall buildings require careful planning for mounting. Wind loading and thermal response create appreciable building sway above approximately 20 stories in height. Beam dispersion and active tracking mechanisms can be used to mitigate misaligning in these situations.

Another issue to consider is that most equipment will not function if the disk of the sun imposes itself directly behind the link head. Saturation of the opposite photo diode receive area occurs and the link will go down for up to several minutes for the few days per year that this solar position occurs. To eliminate the potential for this problem, either avoid an east-to-west orientation, or position the link heads so that the building or some another barrier shadows them.

Installation of Free-Space Optical Systems

CHAPTER 6

129

6

INSTALLATION OF
FREE-SPACE
OPTICAL SYSTEMS

Power Considerations

After the system is mounted, a certain amount of electrical work must be done to ensure proper power supply to the link heads.

Each FSO link head comes with a variety of power requirements, including 110 and 220 volts, 50 and 60 Hz AC, as well as 24 and 48 volts DC with both internal and external converters available, depending on manufacturer. However, the service providers might have their own power requirements. It is important to understand those requirements and ensure such power supply.

In most cases, a dedicated circuit is preferred to reduce inadvertent circuit interruptions, and an uninterruptible power supply (UPS), surge protection, ground fault interrupt circuit, and grounding and lightning protection are recommended. See local codes for specific local code requirements, and manufacturers' documentation for specific model requirements and recommendations.

The power outlet should provide at least one open receptacle for powering diagnostic equipment, power tools, and so on. Many link heads are supplied with a conventional power cord that can simply be plugged into a grounded receptacle with a weather-tight door.

Cabling Considerations

Depending on the specific equipment's network interface, cabling can be either copper UTP (unshielded twisted pair), CAT3 (RJ45 connector), copper UTP CAT5 (RJ45 connector), BNC coaxial, or fiber-optic cable (SC or ST connectors/multimode or single mode). Typically, T1-E1 models incorporate copper UTP CAT3 or coaxial cabling; 10 Mbps Ethernet models incorporate copper UTP or fiber-optic cable; and most others incorporate a fiber-optic cable interface. Fiber optic cable offers (at higher cost) the advantages of longer segments, higher bandwidth headroom, and elimination of the lightning protection required for UTP and coaxial cabling. If the network access device (hub, switch, or router) lacks an appropriate available fiber-optic interface, a module will have to be added to each end. If no fiber module is available for the access device, it will either have to be replaced or a copper UTP link head interface selected.

The two greatest potential problems with an installation, assuming that a thorough site survey has properly defined the deployment, are the cabling and the link head "lock down" following final alignment. To eliminate the former, if anyone other than the FSO contractor provides the cabling, that person should be a certified expert in laying cable. If the FSO contractor installs the cabling, it must be verified either by a cable certification, or by a link throughput test as described in the "Verifying the Link" section that follows later in this chapter.

Any rooftop cable that runs must either be an "armored" outdoor rated type or in conduit. All lightening protection and junction boxes must be weather tight and mounted above any potential rooftop water levels. Due to negligible incremental cost, six-strand fiber-optic cable is pre-ferred to two-strand cable for redundancy in the event of a cable break as well as for link man-agement, diagnostics interface deployments, and so on.

Deployment Configuration

Because FSO equipment can be deployed in different configurations such as point to point, mesh, ring, or point to multipoint, it is important that as a part of the site survey, any factors that are pertinent to such deployment be considered.

Infrastructure Installation

After you have completed the site survey, you are ready to prepare for installation of link heads.

> **NOTE**
>
> Each free-space optics manufacturer provides detailed, model-specific installation instructions. These contain important safety warnings (see Figure 6.4) and other criti-cal instructions that an installer must carefully follow. The manufacturer's instructions supercede the following more general information, and should be complied with for safe and effective installation of equipment.

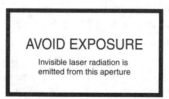

FIGURE 6.4
Typical warning labels that accompany FSO equipment.

Installation of Free-Space Optical Systems

CHAPTER 6

131

6

INSTALLATION OF
FREE-SPACE
OPTICAL SYSTEMS

The installation of the free-space optical link begins with the gathering together of all of the required tools (reference manufacturers' installation manual) and the equipment to be installed. It is recommended that a "rooftop kit" be assembled in a backpack for ease of transport to the rooftop (the most common mounting site). (A little extra time spent on putting together a complete, but not excessive, installation kit will save wasted effort and trips up and down the ladder later!)

Prior to link head installation, verify that line power is in and tested, cabling is in and tested, any required network access modules have been implemented in the access devices (hub, switch, or router), and that the link heads power up like they are supposed to.

Mounting the Link Heads

Installation of the mounts can be a challenging task depending on the chosen mount location. The mounts must be attached to the supporting structure, taking advantage of all available mounting holes (often six or more). Use only stainless steel hardware for strength and to avoid corrosion.

If you must mount to a nonrigid surface (such as wood, sheet steel, and so on), deploy a generous-sized subplate between the mount and the surface to distribute the mount. Tie into subsurface members whenever possible. Remember: The link head mount must be "rock solid" to comply with the rigidity required for accurate aiming a good distance away.

Attachment of the link head to the mount should also be accomplished with stainless steel hardware. Loosen all link head positioning "lock downs" for the alignment process after the link head is in place.

Installing Cabling and Power

Attach the link head to the cabling. This can be done in a separate junction box, or a lightning arrester enclosure (copper UTP cabling).

Connect the link head to the line power supply and power up, observing all cautions noted in the manufacturer's documentation.

Safety

Observe several cautions while installing an FSO link:

If working on a ladder, use extreme caution when hammer-drilling into masonry or steel structures to avoid accidents. Always use a ladder stabilizer. Don't exceed your capabilities; always work in teams of at least two people, carry a complete first-aid kit, and never look directly into the aperture of the link head when laser(s) is active.

Alignment

To ensure the proper functioning of an FSO link, the systems must be properly aligned. In general, one person can perform the installation of a short-distance FSO system. However, for longer-range links, it is recommended that two people (one at both link head sites) perform the alignment to reduce the time to perform the installation process.

The alignment process varies for specific equipment, but it is generally a process of accurately aiming each of the two link heads at one another to enable optical connectivity between them. Alignment is a two-step process.

The first step is coarse alignment. The purpose of coarse alignment is to have the link heads point at each other before the power is turned on. Coarse alignment is obtained by assisted (telescope or internal video camera) or unassisted approximate pointing of the two link heads toward each other. After the link heads are coarsely aligned, turn on the power.

The second step is fine adjustment. After the link heads are switched on, both stations will transmit an optical idle signal. Thread thumbscrews and some type of signal feedback indicator are used to zero in on an optimal alignment. Various aids are incorporated to facilitate the process. At a minimum, an LED digital readout is provided to give feedback as to the strength of the received signal at each end. This is commonly referred to as Received Signal Strength Indicated (RSSI). Some models provide a laptop or PalmPilot interface that is capable of RSSI feedback plus more sophisticated operations, including on-off for squelch, on-off for individual lasers (in multibeam systems), loopback tests, internal temperatures, attenuators (for short distances to prevent saturation), various diagnostic voltage readings, and so on. Still others incorporate simple "dip-switches" to accomplish several of these functions.

It is important to note that some link head RSSI indicators are reading the unit's photo diode receiver; therefore, they are measuring the opposite link head's alignment, not the unit's own outgoing beam aiming. The indicator in this case is affected less by its own alignment than by the alignment of the opposing unit's beam. In this case, alignment is a two-person task, and the opposing unit's RSSI is the measure of a link head's proper alignment. Refer to the specific model's documentation for instructions on interpreting RSSI during the alignment process.

Equipment that is capable of longer distances often includes at a minimum a spotting scope (fitted with infrared rejecting lens coatings for safety) for rough initial alignment. Some models include one or more integrated video cameras that can greatly facilitate aiming on long links.

When using telescopes for alignment, please refer to the laser safety instructions by vendor for the minimum safe distance for use of the built-in telescope.

Installation of Free-Space Optical Systems

CHAPTER 6

133

6

INSTALLATION OF
FREE-SPACE
OPTICAL SYSTEMS

Most link heads incorporate fine-thread thumbscrews (see Figure 6.5) that are used to obtain final alignment. It is extremely important to "lock down" all positioning screws tightly when final alignment is obtained.

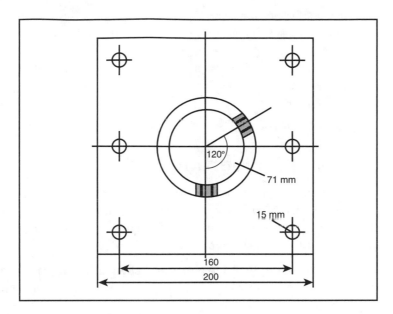

FIGURE 6.5

Fine thumbscrews.

Connection to the Network Interface

After the systems are completely aligned, the last step in the installation and turn-up of the FSO systems is connection to the network. To perform this, connect the optical fibers to the data network interface.

Verifying the Link

Verification of the link at the completion of its installation is the final and necessary step to ensure its integrity. Verification can be accomplished in two ways: bit error rate (BER) and file transfer.

The BER device method is similar to cable certification in that it requires two pieces of equipment: a loopback device at the remote link end (or a loopback setting in the link head) and a bit error rate tester. The BER device generates the desired traffic across the link and measures the bit error rate (which must be less than 10^{-9}) sustained. For low error rates, this test can often be left running for hours—if not sometimes days—to obtain a large enough sampling period.

The file transfer method does not require special equipment; rather, it is accomplished by simply transferring a file of known size (typically approximately 20–100 MB depending on link speed) across the link and timing it. From this information, the performance of the link can be calculated. Allowing for normal operating system or protocol-related overhead, the calculated throughput must be greater than 80% of the maximum-rated throughput of the link (remember to convert the file size from MB [megabytes] to megabits).

Assembly and installation are now complete.

Maintaining and Supporting the System

Most of the free-space optical communications systems are maintenance free. They are designed to handle diverse weather conditions of variable temperatures. The design of the systems makes them durable and not prone to failures other than the normal wear and tear with time. The MTBF of such systems ranges from 15–23 years.

The maintenance of the link head is limited to semiannual cleaning of the heatable front window. Use only a soft, moist cloth and water for cleaning, and follow vendor recommendations on maintenance.

Do not open the housing of the FSO systems. Doing so could be dangerous and could void the warranty.

Extreme care must be exercised when standing close to or touching the link head. It takes very little motion to cause the units to become misaligned.

Most manufacturers' warranties provide for depot repair of defects during the first year; some manufacturers offer three-year standard warranties with extended warranties as well. Terms of warranties vary with the vendors. Repair scenarios could include sending the link heads in for repairs and having a temporary link-up, or participating in an "advance exchange" program that would replace the damaged link head for a fee. In some cases, local partners as well as those who are trained could provide 24×7 services. Another possibility is to stock a few extra links.

Like any other systems, FSO systems can undergo failure. Although most FSO systems are maintenance free and have robust designs, a service provider or end user must be prepared for failure. Figure 6.6 is a sample flowchart used for troubleshooting in the event of a system failure.

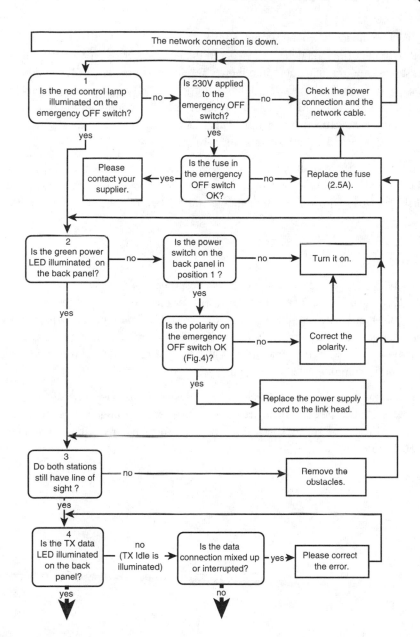

FIGURE 6.6

Troubleshooting flowchart.

Although reactive modes of troubleshooting exist, one can also proactively troubleshoot. That is achieved by monitoring and detecting errors in the FSO systems. Free-space optical communication systems are often integrated into more complex networks to fulfill mission-critical networking tasks. To facilitate the network administrator's task, FSO equipment typically either incorporates Simple Network Management Protocol (SNMP)–compliant manageability, or offers an optional management application. This functionality can provide alarms for link failures and monitor other critical information in real time.

If the failure is to be found in the power supply unit, remember that only authorized technical personnel can conduct checks of the emergency OFF switch and fuse. In all cases, the system must be disconnected in advance from the power supply before engaging in a troubleshooting exercise.

Summary

The site survey is an essential task that is directly responsible for ensuring successful installation and deployment of an FSO system. All elements of the site survey—such as link distances, mounting requirements, cabling requirements, line of sight, deployment configurations, and power requirements—should be properly documented and conveyed to both parties. Additionally, any other activities necessary for deployment—such as obtaining roof rights and constructions—should be properly communicated to avoid delays.

Successful installation and support of free-space optics equipment is straightforward. Because the technologies behind the equipment are quite mature, the learning curve is short, training is easy, and the end users are self reliant and can deploy after they are trained, and maintain and support their FSO investment.

Free-Space Optics and Laser Safety

IN THIS CHAPTER

Laser safety and the proper use of lasers have been a source of discussion and standardization efforts since the devices first began appearing in laboratories more than two decades ago. The two major concerns are human exposure to laser beams and the use of high voltages within the laser systems and their power supplies. Several standards have been developed covering the performance of laser equipment and the safe use of lasers. Tabulations of these standards by industry and government agencies are available (Weiner, 1990).

This chapter covers laser safety with an emphasis on eye safety, as well as the most recent U.S. standards for classifying lasers.

Lasers and Eyes

Certain high-power laser beams used for medical procedures can damage human skin, but the part of the human body most susceptible to lasers is the eye. Like sunlight, laser light travels in parallel rays. The human eye focuses such light to a point on the retina, the layer of cells that responds to light. Like staring directly into the sun, exposure to a laser beam of sufficient power can cause permanent eye injury.

For that reason, potential eye hazards have attracted considerable attention from standards writers and regulators. The standards rely on parameters such as laser wavelength, average power over long intervals, peak power in a pulse, beam intensity, and proximity to the laser.

Laser wavelength is important because only certain wavelengths—between about 400 nm and 1,550 nm—can penetrate the eye with enough intensity to damage the retina. The amount of power the eye can safely tolerate varies with wavelength. This is dominated by the absorption of light by water (the primary component in the eye) at different wavelengths. Figure 7.1 shows the eye's response to different wavelengths. The solid line reflects the visible region and the dashed line shows the total response across near-infrared wavelengths.

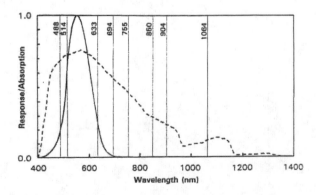

Figure 7.1

Absorption versus wavelength in the human eye.

The vitreous fluid of the eye is transparent to wavelengths of 400—1,400 nm. Thus, the focusing capability of the eye causes approximately a 100,000-to-1 concentration of the power to be focused on a small spot of the retina. However, in the far infrared (1,400 nm and higher), such light is not transmitted by the vitreous fluid, so the power is less likely to be transferred to the retina. Although damage to the corneal surface is a possibility, the focusing capabilities of the eye do not lead to large magnification of power densities. Therefore, much greater power is required to cause damage. The relevance of this is that lasers deployed in FSO that utilize wavelengths greater than 1,400 nm are allowed to be approximately 100 times as powerful as FSO equipment operating at 850 nm and still be considered eye safe. This would be the "killer app" of FSO except that the photo diode receiver technologies suffer reduced sensitivity at greater than 1,400 nm, giving back a substantial portion of the gain. Also, lasers that operate at such wavelengths are more costly and less available. Nevertheless, at least one FSO manufacturer has overcome these obstacles and currently offers equipment deploying multiple 1,550 nm lasers.

With respect to infrared radiation, the absorption coefficient at the front part of the eye is much higher for longer wavelength (> 1,400 nm) than for shorter wavelength. As such, damage from the ultraviolet radiation of sunlight is more likely than from long wavelength infrared. Eye response also differs within the range that penetrates the eyeball (400 nm—1,400 nm) because the eye has a natural aversion response that makes it turn away from a bright visible light, a response that is not triggered by an (invisible) infrared wavelength longer than 0.7 μm.

Infrared light can also damage the surface of the eye, although the damage threshold is higher than that for ultraviolet light.

High-power laser pulses pose dangers different from those of lower-power continuous beams. A single high-power pulse lasting less than a microsecond can cause permanent damage if it enters the eye. A low-power beam presents danger only for longer-term exposure. Distance reduces laser power density, thus decreasing the potential for eye hazards.

Laser Safety Regulations

Many countries have safety standards that must be met by laser products sold there. The National Center for Devices and Radiological Health (CDRH)—part of the U.S. Food and Drug Administration (FDA)—has established standards in the United States (CDRH, 1985). Many other countries have individual standards based largely on recommendations of the International Electrotechnical Commission (IEC, 1984). Laser product standards include requirements for warning labels indicating laser beam classification and type. The higher the classification, the greater the potential hazard to humans. In the U.S., a Class IV identification represents the most powerful lasers. Depending on hazard classification, lasers sold in the United States might require beam shutters to block the beam when not in use, key interlocks, and other safety features.

The United States does not have federal standards for use of lasers, but several states have set their own standards. In addition, many other countries have standards for laser use. The IEC-recommended standards cover the safe use of lasers. The American National Standards Institute (ANSI) has developed voluntary standards for laser use. These standards recommend avoidance of eye exposure.

Because various organizations participating in laser safety have developed slightly different standards and classification schemes, the IEC and FDA sought to develop a unified standard to cover the use of laser systems internationally. This effort was driven by the idea of global markets, and the IEC took the initiative to amend/modify the international IEC 60825-1 standard. The IEC adopted this new classification standard (IEC 60825-1 amendment 2), effective March 1, 2001, in all countries covered under IEC regulations. The FDA/CDRH has committed to unifying its compliance standards with those established by the IEC in the near future. Compliance with the FDA/CDRH laser power standards prior to March 1, 2001 is still in effect during this interim period pending a filed and accepted variance with the FDA/CDRH. In this chapter, we will briefly discuss some of the differences between the previous IEC and the FDA/CDRH standards. However, the primary focus of this chapter will be on the new IEC standard definition and how it relates to free-space optical products.

Comprehensive coverage of laser safety is beyond the scope of this book. The publications listed at the end of this chapter are good resources for more information. In addition, many of the organizations described in Appendix B, "Laser Safety Resources," publish laser safety standards, some of which are available over the Internet.

Laser Classification

To enforce and ensure safety, IEC and CDRH created classifications for lasers. The (old) IEC 60825-1 standard and the CDRH of the FDA 21 CFR Ch.I, Part 1040.10 standard were not completely aligned; in fact, they showed slightly different classification criteria. In general, both standard/regulatory bodies separated lasers into laser Classes I through IV, based on parameters such as laser wavelength, average power over a specific time interval, peak power in a pulse, beam intensity, and distance from the laser.

Some of the four major laser classes were subdivided into groups characterized by an alphabetical letter, such as Class IIIA and Class IIIB. Although FDA and IEC (under the IEC 60825-1 ruling) had slightly different classification schemes, the nomenclature was similar. As an example of a difference, Class IIIA covers infrared wavelengths according to the IEC 60825-1 regulation, whereas it covered only wavelengths up to 700 nm according to the FDA standard. Both standards also required slightly different protection mechanisms, such as key locks, remote interlocks, or shutters for higher-power laser systems.

Amendment 2 of the IEC 60825-1 regulation unifies these small differences. In general, the new standard positively affects free-space optics manufacturers because it allows for launching higher power levels at lower laser classification levels. The new regulation especially stresses the fact that power in free-space optics systems is launched from extended sources such as a larger-diameter lens—and not from the narrow-diameter emission spot of a typical laser source. Percentage-wise, the increase of the allowed power levels is higher in the shorter wavelength band of the infrared spectrum (around 850 nm) than in the longer 1,550 nm wavelength range.

Within this chapter, we will focus on the lower power laser classification standards because they are the only relevant classes for free-space optical systems. If you follow these guidelines, the free-space optical systems are eye-safe and do not require controlled access by untrained personnel, such as window washers or maintenance workers on roofs.

The following new international laser safety classifications are the most relevant to free-space optics products for guaranteeing safety, according to IEC 60825-1 amendment 2:

- Class I: Lasers that are safe under reasonably foreseeable conditions for operations, including the use of optical instruments for intrabeam viewing.
- Class IM: Lasers emitting in the wavelength range from 302.5 nm to 4,000 nm, which is safe under reasonably foreseeable conditions but might be hazardous if the user employs optics within the beam.

Two unsafe conditions are possible when working with Class IM lasers:

- For diverging beams, if the user places optical components within 100 mm of the source to concentrate (collimate) the beam
- For a collimated beam with the aperture larger than the aperture specified in Table 7.1 for measurements of irradiance and radiant exposure[1]

TABLE 7.1 Laser Power Classification According to IEC and CDRH

	Power (mW)	Aperture Size	Distance (mm)	Remark (mm)	Power Density (mW / cm²)
		Laser Class 850 nm			
CDRH Class I (old)	0.076	7	200	All else IIIB	0.20
		50		With optics	
IEC Class I (old)	0.44	50	100		0.02

TABLE 7.1 Continued

	Power (mW)	Aperture Size	Distance (mm)	Remark (mm)	Power Density (mW / cm^2)
IEC Class IIIA (old)	2.2	50	100		0.11
	0.44	7	100		1.14
IEC/CDRH Class I (new)	0.78	7	14		2.03
	0.78	50	2,000		0.04
IEC/CDRH Class IM (new)	0.78	7	100		2.03
	500	7	14		1,299.88
	500	50	2,000		25.48
IEC/CDRH Class IIIR (new)	3.9	7	14		10.14
Laser Class 1,550 nm					
CDRH Class I (old)	0.79	7	200		2.05
		50		With optics	
IEC Class I (old)	10	50	100		0.51
IEC Class IIIA (old)	50	50	100		2.55
	9.6	3.5	100		99.83
IEC/CDRH Class I (new)	10	7	14		26.00
	10	25	2,000		2.04
IEC/CDRH Class IM (new)	10	3.5	100		103.99
	500	7	14		1,299.88
	500	25	2,000		101.91
IEC/CDRH	50	7	14	129.99	
Class IIIR (new)	50	25	2,000		10.19

[1] *50 mm lens at a distance of 2 meters*

Power Limitations for the New IEC60825-1 (2) Standard

Because power density is one of the factors that affect laser safety, it is important to talk about so you can understand its effects. Table 7.1 shows the power level limitations according to the new IEC 60825-1 amendment 2 standard for Class I and IM laser systems. The upper part of Table 7.1 refers to 850 nm transmission wavelengths, and the lower part depicts the same information for 1,550 nm transmission.

The safe laser power limits and responding power densities—according to the specific classifications determined by the FDA and IEC—are shown in columns 2 and 6, respectively. Column 2 depicts the power level that is allowed at the specified wavelengths for an aperture size given in column 3. In essence, the laser beam is directed at the aperture (normally a plate with a hole in the middle). An optical power meter measures the total power that is emitting through the aperture. The exact distance between the aperture and the laser source (or emitting lens) is determined within the standards document. The specific distance for a given aperture diameter is shown in column 3. As an example, for a wavelength of 850 nm and compliance with the IEC Class IM (new) standard, a device is permitted to have 0.78 mW of total power collected in a 7 mm aperture and located 100 mm away from the transmission aperture. When more than one row exists for a given classification standard, all three measurement criteria must be fulfilled. Figure 7.2 depicts the measurement methods.

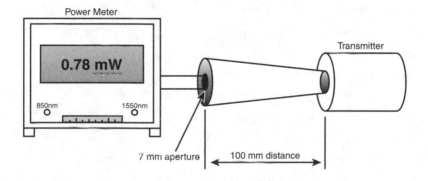

FIGURE 7.2
Laser power measurement IEC Class IM (850 nm).

The new IEC standard allows the output power to increase by a factor of 2 in the 850 nm wavelength range, while still maintaining the eye-safe Class I classification. This is especially important given recent eye-safety discussions in the industry regarding operation at 850 nm or in the 1.5 μm wavelength band. This "power boost" allows manufacturers of shorter wavelength medium speed and medium distance FSO systems (10 Mbps—1,000 Mbps and < 2 km) to increase the system availability by keeping the overall system cost at a low level when compared to systems operating around 1,550 nm. For longer distances and ultra high-rate transmission systems (OC-48 and above), operation in the 1,550 nm wavelength range certainly offers many advantages because of the higher transmission power levels that are allowed for eye-safe operation.

Keep in mind that operation at 1,550 does not guarantee eye safety. Whether a vendor is selling products operating at 850 nm or 1,550 nm, it is important to ensure that their products are Class I or Class IM to ensure proper safety.

Methods to Ensure Eye Safety

In addition to compliance with Class I and Class IM IEC 60825-1 amendment 2 standards, manufacturers are responsible for determining if other measures should be taken to further minimize the danger from exposure to infrared radiation. The following sections describe in greater detail how free-space optical products can ensure safety compliance.

Specifically, compliance is attained by the following:

- Limiting laser output power
- Using multiple transmission sources
- Minimizing access to the laser
- Displaying proper eye safety labels
- Providing visible indication of laser on/off status
- Providing for remote power interlock
- Properly locating system controls
- Using safe alignment procedures
- Training users on proper setup and maintenance procedures

The following sections describe these measures in greater detail, with the exception of limiting laser output power, which was discussed previously.

Using Multiple Transmission Sources

Free-space optical systems that use multiple lasers (typically three or four) allow increased total light intensity for longer distances or higher speed while maintaining safety. With respect to eye safety, this approach minimizes the total power launched from a single transmission lens but keeps the overall total power at a level to maintain highest system availability. The system is designed in such a way that the user will not be able to look into all apertures simultaneously at shorter distances. This approach improves scintillation mitigation in the longer distance links, and so is common at ratings of greater than 1,000 meters.

Minimizing Access to the Laser

Care must be taken during installation to ensure that lasers are mounted in such a way as to restrict access to untrained users. This is most often addressed through a program of user training so that proper installation procedures are followed. Minimizing access to the laser ensures that untrained users do not have casual, unescorted access to laser equipment. This is typically managed through use of locked rooms (in behind-the-window configurations) or restricted access to rooftops. Ensuring safe operation is an important aspect of user training in the operation of free-space optics products.

Labeling

All standard bodies require product labeling that is dependent on the class of operation as specified in standards documentation. These labels should clearly identify the class of operation and the standards body to which it applies. The label also identifies that the device is a laser device and provides its wavelength and maximum output power. In addition, a label should be affixed that shows the full name and address of the manufacturer and the manufacture date. They should be clearly visible on the product and positioned in a way that they do not require an individual to be positioned in a potentially unsafe or exposed location.

Visible Indication of Laser On/Off Status

All free-space optical products should include visible indicators that are lit when the product is powered on and emitting laser power. As with the labels, these indicators should be located such that observing them does not require the operator to be located in an unsafe position. Multiple visual indicators are recommended to provide redundancy if one ever burns out.

Location of Controls

As with the labeling and visual indicators, the control mechanism for the laser device should be positioned in a way that controlling the device does not require a user or operator to be in an unsafe location—one in which he is exposed to laser radiation. It is recommended that free-space optical products have their control mechanism at the rear of the device, 180 degrees from the emitted laser radiation.

Safe Alignment Procedures

Alignment of a two-link laser system might require that a user look in the general direction of a distant laser; thus, it is important to ensure that alignment is not unsafe to the operator. This is accomplished in two ways. In general, the radiation from a free-space optical laser system diverges when leaving the device and the power is dissipated, scattered, and absorbed through the atmosphere. When a laser diverges, it causes the total amount of power to be concentrated over a larger area. The laser safety specification is for concentrated power (power per unit area) rather than absolute power. Second, some of the power is lost due to absorption and scattering through the atmosphere. Because most FSO products require a minimum separation between laser links to ensure proper operation (typically more than 20 meters), the user will not be forced into a potentially unsafe condition. (This factor is also a part of the "User Training" section.)

Finally, it should be noted that field glasses and spotting scopes extend the unsafe distance. Many FSO link heads include a spotting scope that is used in the positioning and alignment process. These scopes must have lens filters that reject the near infrared light produced by the laser being viewed so that they do not create an unsafe condition. On some higher-powered devices, specialized internal video cameras are deployed, which provide locating and aligning functionality, virtually eliminating the use of spotting scopes entirely.

User Training

A final imperative to a laser safety compliance program is that the device manufacturer provides an effective user training program. The previously mentioned measures ensure that innocent, passive activities do not introduce unsafe conditions.

For every product sold, a vendor should provide a comprehensive set of product manuals that provide all necessary user instruction in the safe operation of its FSO products. A key portion of this training involves educating the device users about properly placing the laser equipment. Three location types are to be considered: controlled, restricted, and unrestricted.

Restricted access areas allow laser Classes I, IM, II, IIM, and IIIR to be installed. Controlled access areas permit laser Classes I, IM, II, IIM, IIIR, IIIB, and IV to be installed. The primary difference between these two categories is that in a controlled area, access to the installation area is permissible to those who are authorized with laser safety training. Restricted access areas are normally inaccessible to the general public, but accessible to other personnel who might not have laser safety training.

Summary

Proper compliance of all the standards and classifications combined with effective product design and thorough user training will ensure that the lasers and LEDs used in free-space optics equipment are safe.

Bibliography

Hecht, Jeff. *Understanding Lasers*, Sams Publishing, Indianapolis, 1988. (Tutorial introduction to lasers.)

Laser Focus World Buyers' Guide, PennWell Publishing, P.O. Box 989, Westford, MA 01886. Write publisher for information.

Lasers and Optronics (staff report), "*The Laser Marketplace-Forecast 1990*," Lasers and Optronics 9 (1):39–57 (1990). (Annual market overview.)

Lawrence Livermore National Laboratory: *A Guide to Eyewear for Protection from Laser Light*, Lawrence Livermore National Laboratory, Livermore, California, 1987.

Performance Standards for Laser Products, National Center for Devices and Radiological Health, Publication No. HFX-430 (federal standard). To obtain a copy, write NCDRH, 1390 Piccard Dr., Rockville, MD 20850 (standard).

Photonics Directory, Laurin Publishing Co., Berkshire Common, P.O. Box 1146, Pittsfield, MA 01202. Write publisher for information.

Sliney, David and Myron Wolbarsht: *Safety with Lasers and Other Optical Sources*. Plenum, New York, 1980. (Exhaustive review of laser safety, totaling more than 1,000 pages.)

Weber, Marvin J. (ed.): *CRC Handbook of Laser Science and Technology*, 2 vols., CRC Press, Boca Raton, Florida, 1982; also Marvin J. Weber (ed): *CRC Handbook of Laser Science and Technology Supplement 1*, CRC Press, 1989.

Weiner, Robert: "*Status of Laser Safety Requirements*," Laser and Optronics 1990 Buying Guide, 327–329, 1990. (Tabulation of standards documents, updated in each annual edition.)

Winburn, D.C., *Practical Laser Safety*, 2nd ed., Marcel Dekker, New York, 1990. (Practical guide to laser safety, by former laser safety officer at the Los Alamos National Laboratory.)

Service Provider Issues

IN THIS CHAPTER

The service provider industry presents a substantial new opportunity for deployment of free-space optics. The announcement of a major U.S. ILEC in the third quarter 2001 that an FSO vendor's products have passed rigorous trials signaled the beginning of a more mainstream adoption of this technology to many observers of the FSO market space. It is not that the carrier segment is allergic to new technologies; the delays are primarily due to the stringent measures that carriers undertake in approving a technology.

In this chapter, you will learn about some of the unique requirements of service providers, and how FSO attributes map to them.

The Shift to Carrier Class

The telecommunications industry is being revolutionized by a constant shift in bandwidth needs. As the Internet increasingly plays a major role in business-to-business and business-to-consumer e-Commerce as well as most other aspects of life, the need for high-speed connectivity grows. These needs have generated a significant broadband connectivity bottleneck in metropolitan environments around the world. The solution that addresses this need is the build-out of extensive high-performance metro optical networks.

As more and more users at the edge require high-bandwidth connectivity and as more and more service providers are finding themselves stranded in infrastructure-related "still waters," another shift is happening at the optical layer. This shift will drastically impact the medium of connectivity.

While these growing networks face a massive imbalance in infrastructure deployment due to costs, availability, limitations, and speed, service providers are looking at alternative mediums to address this shifting bottleneck. This shift is driving free-space optics to move from only being viewed as an enterprise technology for point-to-point solutions to a mainstream carrier class solution. As a result of this shift, FSO is beginning to be deployed for core applications such as metropolitan network extensions, SONET ring closures, cellular network extensions, network redundancy, wireless backhaul, gigabit Ethernet access, lambda extensions, and backup links.

Given the ongoing bandwidth revolution, FSO vendors are developing products with service provider requirements in mind. These requirements range from providing deployment logistics and 5-9 reliabilities to optimized costs, increase of service velocity, and expedited generation of carrier revenues. In short, it is an all-encompassing and demanding set of requirements.

Carrier class as a descriptive term means different things to different people. For the purposes of this discussion, the term implies a certain set of features that a service provider—that is, a carrier—can "command" before making any decisions to lease or buy an FSO vendor's or any other products. Simply put, these are features that enable a carrier to build and deploy a network consistent with its service level agreements (SLAs). These features revolve around providing a level of service that not only adds value to a network but also ensures a certain level of communication standards.

Characteristics of Carrier Class FSO

The features that any FSO vendor should focus on to establish carrier-class performance using free-space optical (FSO) systems include the following:

- Availability and the coveted 5-9s
- Multiprotocol support
- Optical transparency
- Distance and bandwidth
- Service level agreements (SLAs) that address their availability needs
- Flexibility in deployment (roof and windows)
- Build as they come
- Flexible topology (point-to-point, mesh, ring)
- Network management
- FSO network planning
- Costs
- Seamless integration
- Service velocity

Availability and the Coveted 5-9s

I am sure that at some point in time, you have heard of the term *5-9s*. This translates into 99.999% availability. To further refine this concept, 99.999% availability means that a service provider's network will not be down for more than 5 minutes a year. That is quite a challenge. The origin of this goal comes from voice networks, in particular the 911 emergency call system that requires the network to be up all the time.

In practice, networks are never 100% available, nor are the excessive added costs of the redundancies required to target 100% availability usually tolerated. Rather, availability is always a compromise between cost and benefit as well as your overall network design. Free-space optics links can now be designed to provide virtually any requested level of availability. It's just a matter of cost.

Availability is not a function of the link itself, but of the network design. Even a fiber-optic network that is designed with no redundancy cannot offer 99.999% availability. Because FSO is an open-air medium transmission technology, it is affected by environmental conditions that in turn impact the availability. As discussed in Chapter 3, "Factors Affecting FSO," these conditions include scintillation, absorption, scattering, beam spread and wander, and others. Even though these issues affect the availability of a free-space optical link, they have solutions.

Scintillation

Recall from Chapter 3 that scintillation refers to the blurring of light waves caused by heat. This effect can impact availability of the system. You can address this scintillation in multiple ways. One way is a multibeam approach. The multibeam approach includes transmitting the same information over multiple beams that have separate paths. This is known as *spatial diversity*. Most metropolitan deployments are not affected by scintillation because the deployments are high above the ground.

One important observation is that the farther you move away from the source of scintillation, the smaller the scintillation pockets (small regions where the refractive index of the air is diverse due to temperature differentials). For this reason, a multibeam system is able to combat the effects of scintillation. Statistically, it is not probable that the same beam will hit the same pocket. Therefore, scintillation becomes less of a factor with the multibeam spatially diverse system.

Absorption

Absorption is caused by the beam's photons colliding with various finely dispersed liquid and solid particles in the air such as water vapor, dust, ice, and organic molecules. The aerosols that have the most absorption potential at infrared wavelengths include water, O_2, ozone, and CO_2. Absorption has the effect of reducing link margin, and therefore distance.

Absorption simply means that the photons in the free-space communication path are lost to water molecules present in the atmosphere. Loss of these photons directly impacts the transmission distance as well as the availability of the link. To address this very common atmospheric effect, adjustment of power levels is required. Another factor impacting absorption is the wavelength. So to maintain the same link margins (hence the same distances) one is required to increase the power of transmission of 1,550 nm-based FSO products more than 850 nm-based products.

Scattering

Scattering simply means that when light collides with any particles present in the atmosphere, it undergoes a change in its path. Scattering, depending on the ratio of the wavelength to the size of the particles, is classified as either Rayleigh or Mie. As discussed in Chapter 3, Rayleigh scattering occurs when the wavelength is much larger than the scattering particle. It is not usually a significant factor in FSO. Mie scattering occurs when the wavelength is of the same size as the scattering particle. This is the type that affects the beam of an FSO link.

> **NOTE**
>
> Keep in mind that environmental factors are not homogenous across all areas. For instance, scattering might be nearly nonexistent in one area due to cleaner air, but significant in an industrial section of town.

Refractive Turbulence

Refractive turbulence is a by-product of the presence of turbulent eddies in the atmosphere. Refractive turbulence causes two main effects:

- Beam wander is caused by turbulent eddies that are larger than the beam.
- Beam spreading is the spread of an optical beam as it propagates through the atmosphere.

The effects of both beam spreading and beam wander is that they negatively impact the link performance. Beam wander reduces the link margin by deflecting the beam from its original propagation direction. This causes the loss of photons at the receiver side, and in the worst case, the beam completely misses the receiving terminal and therefore causes a total loss of the signal. Beam spreading causes the beam to spread out over a larger area and therefore the total power of the receive signal decreases. Consequently, this can lead to an increased value of the BER in case the system operates close to the detection limit of the receiver. To address these environmental factors, beam tracking can be very helpful. This will of course only work if the tracking system is faster than the beam wander frequency.

To address these environmental factors, incorporating an active tracking mechanism is highly effective. This active tracking mechanism addresses beam wander by operating at a frequency faster than the beam wander frequency and addresses beam spread by keeping the beam tightly collimated.

Movements or Sway

Buildings move—especially high-rise buildings. It is not intuitive to think that stationary buildings move, but they do. These building move due to sway that is caused by high winds, the thermal response of their materials, and seismic activity. The higher you go in the building, the more sway is present. Both sway and seismic activity affect the alignment of the FSO systems; therefore, they directly impact the availability of the link.

The good news is that sway and other movements such as seismic activities can be addressed such that they will not cause misalignment to reduce availability. Two primary methods to address sway and seismic activities are beam divergence and active tracking.

Beam divergence is the most economical method. It allows the beam to spread such that the diameter of the diverged beam mitigates any potential misalignment. An example would be a beam with 3 milliradians (mrd) of divergence, which means that at 1 km, the beam has a diameter of 3 m. Depending on the potential for misalignment, this could be enough to ensure that the beam is effectively aligned at all times. You can also choose a beam with 6 mrd of divergence. The disadvantage of this approach is that the higher you go in divergence, the shorter you go on distance, which decreases your link margins. This leads to the next method.

Active tracking, a more expensive and sophisticated method, is a proactive approach to address misalignment caused by movements. An important point to remember when selecting an active tracking system is to ensure that it is able to respond faster than the frequency of the misaligning factor (for example, seismic activity); otherwise, you might experience momentary data loss. Ethernet and data-only links are less sensitive to this issue than ATM and voice or video deployments. The active tracking mechanism should be such that it is able to maintain lock of the link at all times and in varying circumstances. No doubt, future FSO equipment designs might incorporate low-cost, solid-state, active-aiming systems that utilize mirrors or optics to maintain perfect aim of a single beam with virtually zero divergence.

Multiprotocol Support

Metropolitan optical networks are characterized by a melange of protocols, and the ability to handle such multiple protocols from a single platform is a requirement for ease of deployment and wider acceptance of FSO links. The alternative—deploying and maintaining multiple, distinct metro-area protocol infrastructures—clearly is an unfeasible proposition for service providers who need to provision services quickly under increasingly dynamic, unpredictable customer requirements. Such a model is prohibited by factors like maintenance and deployment costs along with lengthy right-of-way concerns (for expansion using fiber). Therefore, multiprotocol and multiservice support on a single, common platform and fiber infrastructure is of paramount importance to metropolitan service providers.

Moreover, multiprotocol internetworking capabilities will allow service providers to leverage off of existing fiber infrastructures. The advantages to such a model are multifold:

- It allows the service providers to maintain backward compatibility (for example, SONET-over-DWDM).
- It yields significant cost reductions (by eliminating layers and equipment).
- It offers simplified management because the network elements are considerably reduced.

- It addresses co-location issues.

- It has many evolutionary advantages (for example, migration from point-to-point to meshed and even ring infrastructure).

Multiprotocol support is the cornerstone of metropolitan optical networks because it gives service providers an advantage that clearly differentiates them from their competitors. An approach where free-space optical equipment operates at the physical layer, Layer 1 in the OSI model, is optimal. This choice implies that no switching or routing is performed within the FSO link heads enabling the product to work with any protocol (SONET, ATM, IP, and so on).

Optical Transparency

Closely tied to the issue of multiprotocol support is that of signal format transparency—the property that allows a transmission system to accept and deliver information that is unchanged in form or content from input to output (albeit more at the physical layer). Optical transparency (overall protocol and signaling agnostics) has long been claimed as a strong advantage of any type of optical networking technology, be it long-haul, metropolitan core, or access. Given the diverse mix of data-signaling formats at the access side, this capability is crucial in isolating service providers from the constant evolution of newer data formatting standards.

Optical transparency allows service providers to address many issues and translate them into competitive advantages. It reduces channel latency and does not require expensive transceivers (for O-E conversion), offering significant scalability improvements and cost reduction. Because metropolitan area distances are much shorter than in long-haul domains, format transparency will be less susceptible to distance-related problems, thereby making its deployment much more feasible. Considering this important distinction, metropolitan service providers will likely begin to demand transparent networking solutions. Optical transparency is just not a requirement of metropolitan networks, but it is clearly one of the factors that will make metropolitan networks simple and efficient.

Distance and Bandwidth

For a service provider, it is important to be able to address multiple applications in a combination of varying distances and bandwidth availability. Because metropolitan networks are a mix of varied customers with varied needs located at varying distances from a communications point, it becomes paramount for service providers to have a suite of products that addresses these needs. Therefore, the FSO vendors should develop products that address this dynamic and changing traffic pattern in metropolitan networks. For a service provider to be able to use FSO in its networks, it has to have a product line that is able to scale from 10 mbps to 2.5 Gbps and beyond and operate at distances ranging from 50 m to 4,000 m giving service providers enough flexibility to deploy FSO in multiple applications. Again, it is wise to mention that bandwidth and distance are influenced by atmospheric conditions.

Service Level Agreements

At one time, service level agreements (SLAs) were merely a way for service providers to guarantee a certain level or service that they would provide in exchange of fees taken to provide that service. Now, service level agreements are a competitive advantage and a requirement. Most service providers do not sell services without offering an SLA. Therefore, any service provider deploying free-space optical solutions has to be able to satisfy and address the carrier's requirements of the SLA. Most service providers offer SLAs that guarantee 99.9% or 99.7% availability in metropolitan networks. The 5-9s rating means systems will be down for five minutes each year, 4-9s equals downtime of 53 minutes per year, 99.91% equals 473 minutes of downtime each year, and 3-9s means downtime of 526 minutes per year.

Free-space optical systems are able to offer such SLAs, and when properly designed and deployed in optimal weather conditions, they are also able to meet the infamous 5-9s availability. However, the truth is that it is quite unrealistic to expect 99.999% availability from a link without considering all other factors.

Roof Rights

I remember about 10 years ago flying over New York City and seeing rooftops in certain areas that were quite free of communications equipment. More recently, I noticed a stark contrast with rooftops now full of antennas and other telecom-related infrastructure. Rooftops are fast becoming a rather expensive real estate proposition. This might be irrelevant to some service providers, but increasing costs in acquiring and renting rooftop rights for deploying equipment is an issue FSO vendors must consider.

FSO vendors must consider roof rights in developing free-space optical equipment as well. Most vendors have developed equipment that can be deployed on rooftops only or behind windows only. However, a few vendors have developed systems that can be deployed either on rooftops or behind windows or a combination. The flexibility of mounting FSO link heads behind windows in addition to rooftops allows service providers to circumvent situations where roof rights are not feasible due to costs or permits.

Build as They Come

With increasing costs of deploying fiber-based networks and providing services, service providers are faced with several valid approaches:

- Deploy infrastructure before acquiring customers and hope that the sales organization is able to sell the services.
- Lease the infrastructure from a competing player and resell it.
- Acquire customers first and then build.

All three approaches are valid, although each entails risk: The first scenario runs the risk of a sunk infrastructure if the service provider fails to acquire customers who can pay for the build-out or the customer decides to cancel services, leaving the provider with stranded capital. The second choice of leasing services from a competing player might make a service provider uncompetitive from a price-point standpoint and limit the provider's ability to provide acceptable SLAs. The third scenario of acquiring customers before you build simply means loss of revenues and bad customer service. All three scenarios impact the service provider's service velocity, revenues, and market share.

What the service providers need is a platform that allows them to build as they gain customers. The choice should be such that it helps to accelerate their service velocity; that is, it allows them to turn up services quickly and contribute to their revenue growth. Hence, they can start generating revenues right away while gaining market share. Remember that FSO does not involve the significant costs/time requirements of installing wired technologies such as fiber, so it allows service providers to acquire customers first, then build quickly and inexpensively. If the customer decides either to cancel its service or move to a new location, FSO is a technology that can be easily moved to where the customer is. Such a scalable and flexible approach is part of the value proposition of free-space optics.

Flexible Topology

Before the existence of FSO-based networks, there were long-haul networks. More recently, metropolitan networks have been divided into the core, access, and edge. There is an abundance of rings, meshes, and point-to-point links in existing telecommunications networks. From an FSO perspective, what is the network topology that makes sense for the service provider? Before addressing that question, this chapter will review the basic network topologies.

Point-to-Point

Point-to-point systems are connections between two points. These points can be campus buildings, points within a ring, a spur from a ring to a hub, or to connect multiple LANs. Point-to-point systems are simple, cost effective, scalable, and easy to manage and deploy. The systems can easily scale up and are quick and easy to deploy.

Mesh and point-to-multipoint topologies are just special cases of point-to-point. Hence, point-to-point systems can be deployed in any topology. What is different is the additional cost that is added to deploy those topologies.

Mesh

A mesh network, as its name suggests, is a network that is made by interconnecting multiple points in a wide variety of links. Mesh is a special case of point-to-point links in which each point (node) is connected to other points (nodes) in such a way that they form a network that enables a service provider to provide redundancy. Mesh networks are more commonly found in long-haul networks, but they are beginning to move to metropolitan network cores. Although fully meshed networks make sense in the long-haul and core metropolitan networks, their deployment in edge networks is certainly not a viable solution at the present time due to cost and complexity.

Ring and Spur Architecture

Ring and spur architecture is one of the most commonly deployed architectures, and it can be found in the backbone of dense metropolitan networks. Spurs are a cost-effective way for service providers to extend the reach of their networks without the added cost of deploying an additional ring. Service providers typically deploy ring architectures to ensure redundancy and reliability of networks. However, using FSO links, spurs can be used to connect other rings, thus enabling service providers to form an interconnected network at reduced cost.

Point-to-Multipoint

In a point-to-multipoint architecture, a service provider can deploy multiple links from a single point. These multiple links are able to connect multiple points, thus addressing multiple customers. This is an easy method to address the needs of multiple customers from a single source. This topology is most viable for locations where the density of customers is high. Point-to-multipoint architectures are also known as star or hub and spoke.

Multiple Point-to-Point Architecture

As the name suggests, this architecture involves multiple point-to-point connections linked together. Multiple point-to-point architecture addresses the need for long-reach links. It looks like a fiber link and is more economical than deploying fiber in any network. The advantage to this approach to FSO is the ability of a service provider to quickly acquire new customers without waiting to lay fiber and improving their service velocity.

The Optimal Choice: What to Deploy

Of the topologies just described, the most commonly deployed in FSO networks are Layer 1 point-to-point, IP/ATM-based mesh, and point-to-multipoint.

A Layer 1, transparent, point-to-point approach is the most ideal one. This approach is ideal for many reasons, including the following:

- Allows a service provider to transmit any type of information (voice video or data) and use any protocols.

- A service provider is also able to integrate FSO with its existing protocol structure be it ATM, SONET, or IP.

- Provides the service provider with maximum flexibility and freedom to deploy a value-added network.

- Service providers, if needed, can design and engineer the products to work in any network topology, including mesh, point-to-multipoint, ring with spurs, and point-to-point.

The point-to-point approach provides metropolitan service providers the freedom to rapidly build and extend networks that deliver fiber-optic speeds for today's bandwidth-hungry customers.

A mesh approach adds redundancy to a network, but also complexity and cost. The added cost undercuts one of FSO's competitive advantages: its lower cost relative to installing fiber. A mesh approach requires a minimum of three links. A protocol-specific mesh further complicates matters because a service provider is forced to deploy a solution that might not be optimized for its network.

A point-to-multipoint approach assumes that all of the customers are within a line of sight of one building. Availability can be a problem in a point-to-multipoint system because the distances between the multipoint hub and the actual customer side can vary quite a bit. Foliage is a problem as well, but this is a bigger concern in residential areas, which will not be included in the first stage of FSO deployment. For these reasons, a point-to-multipoint topology is less preferred than point-to-point for FSO.

Network Management

Network management is a crucial element of any network. Service providers who have extensive nationwide networks comprised of long-haul, metropolitan, and local area networks conduct their network monitoring and management from a centralized location often referred to as the NOC. NOC stands for networks operation center. The NOC generally is comprised of multiple screens that flash network alarms to NOC operators. FSO manufacturers should ensure that their products are visible to the NOC. This can be accomplished in many ways.

One way is to have an element management system (EMS). An EMS is item specific, which means it is not universal software, but rather allows a service provider to monitor the equipment on which it has been implemented while the FSO network is being deployed and tested. EMSs can be crucial to the large enterprise customers.

The second approach to network management is to have a Simple Network Management Protocol (SNMP) interface integrated into the product. This allows service providers to view the status of various items remotely using an SNMP agent.

Cost

Cost is an important factor for service providers who are considering deploying FSO solutions into their networks. The value proposition of FSO is the cost and bandwidth advantage when compared to other technologies, particularly fiber deployment. This point requires some clarification because many people have an incorrect perception of the cost of fiber.

It is important to remember that only 15% of the cost associated with fiber build-out is the cost of the fiber. The remaining 85% is the cost of the deployment, which includes permits, labor, trenching, and so on. It typically costs between $100,000 and $200,000 to connect a building with fiber to a nearby fiber loop. In some cities, due to construction obstacles, this cost could be millions of dollars. With time, even though the cost of fiber might decrease, the cost of deployment will certainly go up and will become more and more difficult. Some cities are enforcing moratoriums on street trenching to lay fiber-optic lines underground. The rising costs of securing property rights, permits, and labor to perform the digging are increasing rapidly as well.

In contrast, FSO products can offer compelling, competitive, price-point alternatives to fiber. With FSO, service providers do not have an initial high cost because it is license free. In addition, FSO has no sunk cost as is associated with fiber deployment. (After fiber is deployed, it stays in the ground even after the customer leaves.) FSO is an easily redeployable platform that can be moved to where the customer is. Also with FSO, the service velocity is significantly increased, and the cost of installation is significantly low, thereby yielding much better returns. The payback in most FSO deployments is within a year. FSO vendors must be able to provide improved services and performance at lower cost levels to gain strong market acceptance.

With longer range and higher throughput comes higher cost. FSO equipment cost for up to 10 Mbps at 1 km is between $5,000 and $10,000. Systems that provide up to 155 Mbps cost between $10,000 and $30,000, depending on distance, and those that provide 622 Mbps cost between $20,000 and $40,000. Gigabit Ethernet links start between $40,000 and $50,000. Systems are being developed that go higher than gigabit Ethernet, but price points are not being talked about yet; for them to make economic sense, they have to be less than $100,000.

FSO installation is usually one-third of the total cost of the system. That includes site acquisition, labor, preparation, testing, and establishment of connection points to the networks. Key costs are the optical components, housing, and mounting. Vendors expect declines of between 40–50% of current prices in the next three years.

FSO Network Planning

With the increasing adoption of FSO in metropolitan networks, it will become increasingly important to arm the service providers with an FSO planning tool. This tool will allow service providers to plan and design an FSO network that enables them to address the specific needs of a customer while maintaining its SLAs and network requirements. Such a tool will give the service provider a visualized view of a metropolitan area where deployment is being considered, allowing the provider to select buildings, choose distances, and determine line of sight. All these variables are important for a service provider to determine the feasibility of an FSO network in that region.

Seamless Integration into Existing Infrastructure

The metropolitan network is changing, driven by two key factors: customer requirements and technology to support these requirements. As customers move from low bandwidth to high bandwidth and as they switch from one protocol to another, they change platforms, change service, and demand that service providers meet their changing needs. This leaves the service providers with a metropolitan network that is part optical and part electrical, part SONET and part IP, part ATM and part transparent. To further complicate matters, the networks are of varying bandwidths with the core being OC192/10 Gbps, the access being OC48s, and the edge migrating from OC3 to OC12. This is further highlighted by the presence of 1,310 nm and 1,550 nm signals.

The metropolitan networks are a potpourri of protocols, bandwidths, topologies, frequencies, layers, and equipment. For successful adoption and large-scale deployment of FSO in the metropolitan networks, it is important that FSO integrates with existing metropolitan networks. It becomes all the more important that FSO vendors develop products that are transparent so that they can support multiple protocols of variable data rates as well as independent wavelengths so that 1,310 nm and 1,550 nm signals can be transported with the same flexibility.

Service Velocity

Service velocity refers to the time it takes from the moment a request for new or expanded service is received by a provider to the time the service is turned up. This concept is also known as *service provisioning*. The benefit of accelerated service velocity is quicker revenue generation.

It is clear that a connectivity bottleneck exists in metropolitan networks. And as the connectivity bottleneck keeps on growing and shifting around the metro networks, it is clear that a metropolitan "traffic jam" is forthcoming. Service providers who want to participate in this new and large opportunity need to address it quickly. Delay or lack of action will cost the service provider significant loss in revenues and market share. With service velocity as slow as months, bandwidth needs as high as OC48, and fiber deployments slower and increasing in cost, it becomes all the more important for service providers to address this. FSO offers service providers a means to increase the service velocity to days and pulls the revenue streams quickly.

Summary

With so much fiber laid in the ground and the inability of carriers to fill those pipes with traffic due to lack of connectivity to that traffic, carriers are now looking toward FSO to provide that connectivity. Where once the carriers were measured by the fiber miles they had, now they are being gauged by the "bits per route mile." Free-space optics presents a competitive opportunity for carriers, but to deploy FSO in a larger scale throughout their networks, they must ensure that FSO meets the standards they have laid out.

Alternative Access Technologies

IN THIS CHAPTER

This chapter provides an overview of alternative access technologies that are commonly used to connect a subscriber to the network. Many of the newer and higher-speed access technologies have been extensively studied and developed over the past couple of years. These developments were mainly triggered by the need for faster and less expensive Internet access when compared to Frame Relay and leased-line services. For businesses, leased-line services such as T1 (1.544 Mbps), E1 (2.048 Mbps), or higher speed DS-3 (45 Mbps) connections were most commonly used for networking access.

These services originated from a world that was mostly driven by the need for high-quality voice services. The internal framing structure of T1/E1 or DS-3 connections was designed to carry multiple phone calls (channels) at the same time within a guaranteed period of time. This quality-of-service (QoS) feature is important for voice applications because the quality of the voice service suffers without this feature. For Internet applications, these "voice" channels are filled with bits carrying data instead of voice traffic.

Alternative access technologies such as digital subscriber line (DSL) technology or cable modem technology were primarily developed to satisfy the need for data-driven Internet applications. Both xDSL and cable modems are using the existing copper-based local loop infrastructure to connect to the outside network: DSL technology uses standard twisted-pair phone lines, and cable modems use the existing coax cable television broadcast infrastructure. To provision higher speed access, the copper-based infrastructure is the main bottleneck. However, fiber-based infrastructure deployments have been slow due to the high upfront costs and difficulties to lay fiber in densely populated areas.

In addition to these wire-based access technologies, a variety of wireless broadband access strategies was heavily discussed over the past couple of years as alternative routes to bypass the existing copper-based infrastructure. These technologies use the higher frequency unlicensed or licensed microwave transmission bands. Figure 9.1 shows a snapshot of various access technologies that U.S. businesses used for broadband connections in 2001.

The following sections will discuss some of these technologies in more detail and look at them from different angles. Access strategies will be analyzed from a more technical aspect. Other access strategies will be introduced by looking more at the economics from the potential service provider perspective. This approach emphasizes the balance between ease of technical implementation and the economics of a specific access solution. The end of this chapter will explain the role that FSO can play in this arena when compared to other access strategies.

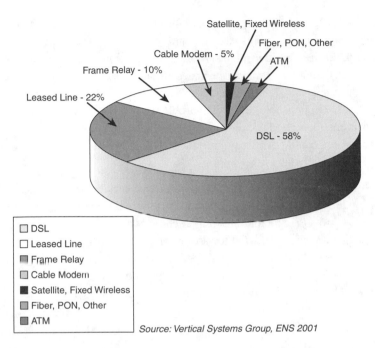

Source: Vertical Systems Group, ENS 2001

FIGURE 9.1

2001 business broadband connections by technology.

Digital Subscriber Lines

There are different "flavors" of DSL, collectively known as xDSL. xDSL technology originates from research done at the Bell Communications Research Laboratory (Bellcore). The original intention of the Bellcore research was to study the use of xDSL technology to provide video-on-demand and interactive TV applications over twisted-pair wires.

The original version of DSL, apart from its later siblings, uses a modem similar to one used for Basic Rate Integrated Services Digital Network (ISDN). This original DSL modem could transmit duplex data—that is, data in both directions simultaneously—at rates of 160 Kbps over twisted-pair copper lines at distances up to 18,000 feet over 24-gauge wire. The multiplexing and demultiplexing of this data stream into two ISDN B (Bearer) channels (64 Kbps each), a D (Data) channel (16 Kbps), and some overhead takes place in attached terminal equipment. By modern standards, DSL does not press transmission thresholds. In its standard implementation (ANSI T1.601 or ITU I.431), it employs echo cancellation to separate the transmit signal from the received signal at both ends. This was a novelty at the time DSL first found its way into the network.

Modern variations of DSL technology provide multiple forms of data, voice, and video to be carried over twisted-pair copper wire. xDSL technology provides the local loop connection between a network service provider's central office (CO) and the customer site. xDSL involves signal-processing techniques to leverage the speed limitations of existing local loop infrastructure to increase the amount of data transmitted over analog lines. This has led it to be touted as one of the most viable options to alleviate the problems of limited bandwidth using the existing wireline copper infrastructure. xDSL technology requires a minimal investment on the carrier's side; therefore, carriers could potentially introduce xDSL services to their customers more quickly and cost effectively than other options.

In theory, a subscriber can choose from a variety of xDSL services that are based on different DSL transmission technologies. However, the distance between the subscriber side and the central office (CO) limits the amount of potential choices between different xDSL services. Different DSL technologies are targeted toward different distances from the CO, as well as different applications. Some DSL technologies operate in a frequency band above baseband telephones service, allowing the subscriber to use the same copper wire for both voice and data service. Some deliver symmetrical bandwidth, and others deliver asymmetrical.

Asymmetric Digital Subscriber Line (ADSL)

As its name implies, ADSL transmits an asymmetric data stream, with much more data transmitted downstream to the subscriber and much less going upstream. The reason for this has not so much to do with transmission technology than with the cable plant. Twisted-pair telephone wires are bundled together in large cables. Fifty pair to a cable is a typical configuration toward the subscriber. However, cables coming out of a central office might have hundreds or even thousands of pairs bundled together. An individual line from a CO to a subscriber is spliced together from many cable sections as they fan out from the central office. As for a typical figure, Bellcore estimates that the average U.S. subscriber line has 22 splices.

Even though Alexander Bell invented twisted-pair wiring to minimize the interference of signals from one cable to another caused by radiation or capacitive coupling (attenuation), this process is not perfect. Signals do couple, especially when frequencies and the length of line increase. Laboratory experiments reveal that sending symmetric signals in many pairs within a cable significantly limits the data rate for a given line length. Therefore, designers sacrificed on upstream bandwidth to add additional downstream.

This tradeoff was made for ADSL because many target applications for digital subscriber services are asymmetrical. Applications such as video-on-demand, home shopping, Internet browsing, and multimedia access feature high data rate demands downstream to the subscriber, but relatively low data rate demands upstream. Depending on distance, ADSL has a range of downstream speeds. Typical values follow, but can vary significantly depending on line conditions and equipment used:

Up to 18,000 feet	< 500 Kbps
16,000 feet	1 Mbps
12,000 feet	3 Mbps
8,000 feet	8 Mbps

Upstream speed is less affected by distance, and maximum speeds can range from 400 Kbps to 1 Mbps. Service providers today typically offer a variety of speed arrangements available at different prices, from a minimum set of typically 384 Kbps down and 128 Kbps up to a maximum set of 8 Mbps down and 1 Mbps up. All of these arrangements operate in a frequency band above plain old telephone system (POTS), leaving the POTS service independent and undisturbed, even if a premise's ADSL modem fails. The typical DSL connection at the subscriber location is shown in Figure 9.2.

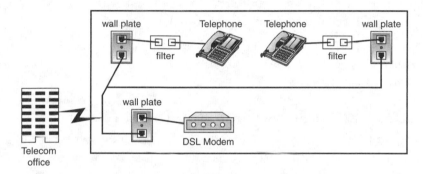

FIGURE 9.2

Typical ADSL wiring at customer premise. The filter allows for simultaneous voice and data operation.

As ADSL transmits digitally compressed video, among other things, it includes error-correction capabilities intended to reduce the effect of impulse noise on video signals. However, error correction introduces about 20 ms of delay, which is too much for certain LAN and IP-based data communications applications, such as two-way packetized voice. Therefore ADSL must know what kind of signals it is passing to know whether to apply error control. The connection of ADSL modems to personal computers and television set top boxes further increase the complexity of the modem. All of these application and operating conditions create a complicated layer of protocols and installation environment for ADSL modems, moving these modems well beyond the functions of simple data transmission and reception equipment. Several large-scale rollouts of xDSL services by carriers have met with failure recently, with implementation complexity cited as a significant contributing factor.

High Data Rate Digital Subscriber Line (HDSL)

HDSL is one of the oldest DSL technologies, and was devised to be a better way of transmitting T1 or E1 signals over twisted-pair copper lines. HDSL uses less bandwidth and requires no repeaters, in contrast to traditional T1 or E1 transmission. Using more advanced modulation techniques, HDSL transmits 1.544 Mbps or 2.048 Mbps in bandwidths ranging from 80 KHz to 240 KHz, depending on the specific technique. In comparison, the bandwidth-hungry Alternate Mark Inversion (AMI) protocol uses 1.5 MHz to accomplish the same task at T1 speeds. HDSL provides such rates over lines up to 12,000 feet in length over two pairs of standard 24-gauge twisted-pair wire.

In the late 1990s, HDSL2 was invented and standardized to further the efficiency of HDSL, taking advantage of improvements in technology to obtain the same data rates (along with improved signal-to-noise ratios) as HDSL, but over a single pair of copper. Both HDSL and HDSL2 operate in a frequency band that includes the band used by POTS, so simultaneous use of the copper for DSL and baseband voice is not possible.

Typical applications of HDSL and HDSL2 include providing T1 services, PBX network connections, cellular antenna stations, digital loop carrier systems, interexchange POPs, Internet servers, and private data networks.

Symmetric Digital Subscriber Line (SDSL)

SDSL was originally just a single line version of HDSL, transmitting half the bandwidth over a single twisted pair. Silicon vendors quickly improved on this technology to offer additional bandwidth, as well as variable rates that could operate over different distances, similar to the techniques employed for ADSL. As with HDSL and HDSL2, SDSL operates in a frequency band that includes the POTS band. These attributes made SDSL technology well suited for business applications, which required symmetrical bandwidth, and typically used multiple phones. This eliminated the need to support voice over the same copper pair as was necessary for the residential market. This gave service providers an alternative to ADSL, and the ability to offer symmetrical data rates above 640 Kbps at longer distances. The standard that emerged was G.SHDSL, and is capable of providing symmetrical bandwidth of up to several Mbps. As with ADSL, depending on distance, G.SHDSL has a range of symmetrical speeds.

Very High Data Rate Digital Subscriber Line (VDSL)

VDSL was originally called VADSL. This is because (at least in its first manifestations) VDSL incorporated asymmetric transceivers at data rates higher than ADSL but over shorter line distances. Although no general standards exist yet for VDSL, discussion has centered around the following downstream speeds (STS-1 corresponds to OC-1):

12.96 Mbps	(1/4 STS-1)	4,500 feet of wire
25.82 Mbps	(1/2 STS-1)	3,000 feet of wire
51.84 Mbps	(STS-1)	1,000 feet of wire

The suggested upstream rates range from 1.6 Mbps–2.3 Mbps. In many ways, VDSL is simpler than ADSL. Shorter lines impose far fewer transmission constraints (such as splices), so the basic transceiver technology is much less complex, even though it is ten times faster. VDSL targets only ATM network architectures, preventing channelization and packet-handling requirements imposed on ADSL. VDSL allows for passive network terminations, enabling more than one VDSL modem to be connected to the same line at a customer premise, in much the same way as extension phones connect to home wiring for POTS.

Even though VDSL offers more bandwidth and its implementation looks easier at first glance, VDSL has quite a few obstacles and limitations. VDSL must still incorporate error correction. In addition, passive network terminations have a host of problems. Some of them are technical, and others are regulatory. This will surely lead to a version of VDSL that looks identical to ADSL, which incorporates inherent active termination. Therefore, the only advantage of VDSL will be its capability for higher data rates. VDSL will operate over POTS and ISDN. Passive filtering will allow separation of these services from the VDSL signals.

Table 9.1 summarizes each DSL technology based on mode, maximum bandwidth, maximum distance over twisted-pair wires, and the types of target applications each technology is best suited for. The table also illustrates the fundamental trade-off between distance and bandwidth of DSL technology. The bandwidth rate decreases as the distance from the customer's site to the CO increases.

NOTE

Mode refers to upstream versus downstream transmission rates. xDSL technologies that are able to transmit data at the same rate both in upstream and downstream directions operate in duplex mode. Technologies that have different transmission rates upstream and downstream operate in asymmetric mode.

9

ALTERNATIVE
ACCESS
TECHNOLOGIES

TABLE 9.1 xDSL Transmission Technologies

Technology	Mode	Bandwidth Capability	Maximum Distance Over 24-Gauge LTP	Target Application
ADSL	Asymmetric	Downstream 1.5–9 Mbps; Upstream 16–640 Kbps	18,000 feet (12,000 feet for speeds above 1.5 Mbps)	Internet/intranet access, video-on-demand, data base access, remote LAN access, interactive multimedia, lifeline phone service
HDSL	Duplex	T1 up to 1.544 Mbps; E1 up to 2.048 Mbps	15,000 feet	Replace local repeated T1/E1 trunk, PBX interconnection
SDSL	Duplex	T1 up to 1.544 Mbps; E1 up to 2.048 Mbps	10,000 feet	Same as HDSL plus premises access for symmetric services, such as videoconferencing
VDSL	Asymmetric	Downstream 13–52 Mbps; Upstream 1.5–2.3 Mbps	1,000 to 4,500 feet	Same as ADSL plus HDTV

Benefits and Limitations of DSL

Despite the fact that a number of DSL technologies exist, most of these are still experimental with no standards or marketplace implementation set in the near future. The one DSL technology that has already been standardized by the American National Standards Institute (ANSI) and introduced successfully into the commercial market is ADSL. Therefore, the discussion of benefits and limitations of DSL will be couched in terms of ADSL specifically.

ADSL's benefits include the following:

- New signal processing techniques are used to leverage existing local loop infrastructure to improve bandwidth capacity of standard analog twisted-pair copper lines.
- ADSL users can use a single twisted pair for both data and voice communications.

ADSL also has limitations:

- Competing standards on how to modulate frequency
- No interoperability due to lack of standards among component manufacturers and carriers
- Cross-talk interference from nearby wires

- Trade-off between length of lines, data speeds, and differences in upstream and downstream traffic

For the consumer, the cost of switching to ADSL service requires the purchase (or lease) of an ADSL modem. If the subscriber wants to use both voice and data communications simultaneously on the same line, a POTS splitter is also required. Unfortunately, in addition to the problem of general availability, the two main reasons that the average consumer has not taken advantage of ADSL are the lack of affordable equipment and service plans. The cost of an ADSL modem is in the hundreds of dollars, and service plans are still relatively expensive. However, more carriers have started to offer ADSL services as part of their package. This will eventually decrease the cost for ADSL service, and the average consumer might find the ADSL option viable from the cost point of view.

For businesses and the SOHO (small office, home office) market, DSL offers a much more cost-competitive solution when compared to ISDN or T1 options. Fifty-eight percent of U.S. businesses use DSL technology as the broadband business connection of choice. However, due to the constraints in distance, the actual business proximity to the CO and whether a carrier offers DSL determine the subscriber's ability to sign up for a DSL service. Although DSL can be a cost-effective access alternative for a business when compared to T1 or ISDN, it will likely never match the amount of bandwidth that can be provided over other access technologies such as fiber, FSO, and wireless.

From the carrier's perspective, DSL costs are spread over three major areas. First, the carrier needs to qualify the line connection between the CO and the subscriber. Some estimates reveal that up to 80% of the twisted-pair lines are viable after some line conditioning. The rest of the lines—the other 20%—are not capable of providing ADSL. However, because most of those wires reside in rural areas, the impact on the total addressable market would be negligible. Second, carriers need to replace their interface cards to be compatible with ADSL line cards. Finally, the carriers need to establish the administrative infrastructure to support ADSL service plans.

Cable Modems

Whereas DSL technology operates through the standard twisted-pair telephone cable infrastructure, cable modems are devices that allow high-speed access to the Internet via a cable television network. Although in some respects, cable modems are similar to a traditional analog modem, a cable modem is significantly more powerful, capable of delivering data approximately 500 times faster. With respect to bandwidth, cable modems and DSL modems are comparable in speed.

Basic Network Cable Concepts

To better understand the nature of Internet access over cable, it is helpful to review how the cable network infrastructure works. The overall architecture or topology of existing residential cable TV networks follows a tree-and-branch architecture, as shown in Figure 9.3. In each community, a "head end," the originating point for cable TV signals, is installed to receive satellite and traditional over-the-air broadcast television signals.

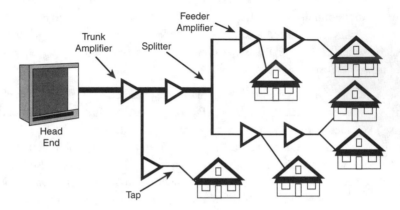

FIGURE 9.3

Coaxial cable tree-and-branch topology.

These signals are then carried to subscribers' homes over coaxial cable that runs from the head end throughout the community. Each 6 MHz TV channel is transmitted in analog form over 6 MHz (not necessarily the same) of enclosed spectrum on the cable. Multiple channels are sent over the same cable using Frequency Division Multiplexing (FDM). Cable modems use 6 MHz bands not being used for television signals to carry data. The speed of this access depends on the equipment used to modulate digital computer information onto cable's analog TV channels. Such equipment typically provides bandwidths ranging from 500 Kbps to more than 10 Mbps.

To achieve geographical coverage of the community, the cables emanating from the head end are split (or "branched") into multiple cables. When the cable is physically split, a portion of the signal power is split off to send down the branch. The signal content, however, is not split but rather travels down the same set of TV channels that reach every subscriber in the community. Figure 9.4 illustrates that the network follows a logical bus architecture. With this architecture, all channels reach every subscriber all the time, whether the subscriber's TV is switched on or not. Just as an ordinary television includes a tuner to select the over-the-air channel the viewer wants to watch, the subscriber's cable equipment includes a tuner to select among all the channels received over the cable.

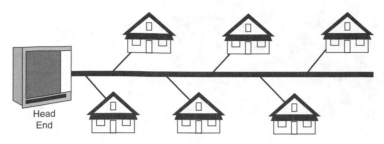

FIGURE 9.4
Logical bus architecture of the cable TV network.

Because the signals attenuate as they travel several miles through the cable to subscribers' homes, amplifiers have to be deployed throughout the plant to restore the signal power. Cables that are split often and that are long require more amplifiers in the plant.

Recent Developments in Cable Architectures

The development of fiber-optic transmission technology has led cable network developers to shift from a purely coaxial tree-and-branch architecture to an approach referred to as Hybrid Fiber and Coax (HFC) networks. Transmission over fiber-optic cable has two main advantages over coaxial cable. First, a wider range of frequencies can be sent over the fiber, increasing the bandwidth available for transmission. Second, signals can be transmitted greater distances without amplification. Reduced cost has been the principal reason that developers have adopted an intermediate Fiber to the Neighborhood (FTTN) approach. However, FTTN approaches are pretty much limited to newly constructed residential neighborhoods because fiber can be connected to each home during the phase of the regular utility buildout.

Figure 9.5 shows a typical FTTN network architecture. Various locations along the existing cable are selected as sites for neighborhood nodes. One or more fiber-optic cables are then run from the head end to each neighborhood node. At the head end, the signal is converted from electrical to optical form and transmitted via laser over the fiber. At the neighborhood node, the signal is received via laser, converted back from optical to electronic form, and transmitted to the subscriber over the neighborhood's coaxial tree and branch network.

In summary, FTTN replaces long coaxial cable runs with long fiber and shorter cable runs. This replacement increases the bandwidth that the plant is capable of carrying. Another advantage is that this approach also reduces both the total number of amplifiers needed and the number of amplifiers cascading between the head end and each subscriber. The total number of amplifiers is an important economic component. Fewer amplifiers and shorter trees also introduce less noise into the cable signal. These improvements translate into higher bandwidth, better quality service, and reduced maintenance and operating expense for the cable network provider.

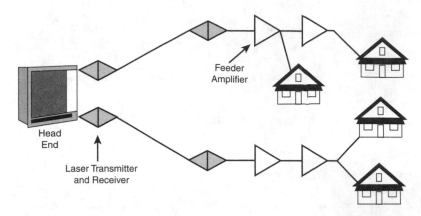

Figure 9.5

Fiber to the Neighborhood (FTTN) network architecture.

Using the broadcast television distribution network is a one-way street because TV programs are broadcast to viewers, but there is no mechanism to send signals back into the network. These systems are designed to send information only downstream toward the subscriber. However, many of the new interactive services require transmission from the subscriber as well, with bandwidth requirements for such upstream transmission varying tremendously depending on the service. Internet applications typically include a range of services that require much upstream bandwidth. This can range from low-bandwidth e-mail, to large e-mail attachments, to high-bandwidth video.

To enable upstream transmission, three types of technical changes to the network are required:

- The spectrum has to be allocated for traffic traveling in the upstream direction. Figure 9.6 shows a typical spectrum map for the signals that are traveling over residential cable plants. Typically, the frequency range from 5–42 MHz is dedicated to upstream transmission. This range generally provides a maximum of four usable upstream channels.

- Amplifiers—including duplex filters—must be included in the plant to separate the upstream and downstream signals and amplify each direction separately and in the correct frequency range.

- Downstream transmission from the head end is broadcasted and the same signal is sent on all the wires and to all subscribers. In contrast, upstream transmission is inherently an individual process because each subscriber is trying to place a different signal onto the network. When going "upsteam," these different signals must eventually share the same piece of transmission spectrum (see Figure 9.7). Therefore, some form of access method is needed to arbitrate which signal is actually carried.

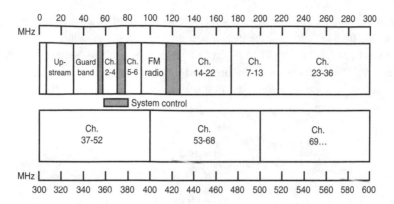

FIGURE 9.6

Cable spectrum map.

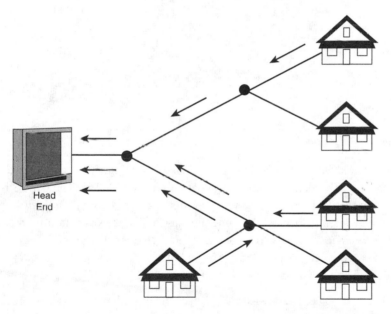

FIGURE 9.7

Sharing of upstream bandwidth.

Like voiceband modems, cable modems modulate and demodulate data signals. However, cable modems must incorporate more functionality suitable for the anticipated high-speed Internet services. From the subscriber perspective, a cable modem represents a 64/256 quadruple amplitude modulation (QAM) RF receiver capable of delivering up to 30–40 Mbps of data in one 6 MHz cable channel, which is approximately 500 times faster than a 56 Kbps modem.

Data from a user to the network is sent in a flexible and programmable system under control of the head end. The data is modulated using a quadruple phase shift keying (QPSK) or 16 QAM transmitter with data rates from 320 Kbps up to 10 Mbps. Similar to a subscriber of a DSL service who can get phone calls at the same time and through the same wire, a cable modem subscriber can continue to receive cable television service while simultaneously receiving data on cable modems (see Figure 9.8).

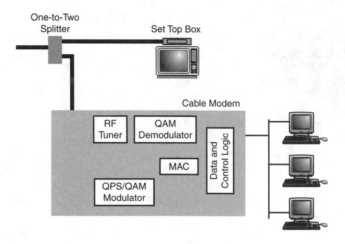

FIGURE 9.8

Cable modem at the subscriber location.

At the cable network head end, data from individual users is filtered by upstream demodulators and further processed by a cable modem termination system (CMTS). The CMTS is a data-switching system that is specifically designed to route data from many cable modem users over a multiplexed network interface. At the same time, a CMTS receives data from the Internet and provides data switching necessary to route data to the cable modem users. Data from the network to a user group is sent to a 64/256 QAM modulator. As a result, user data is modulated into one 6-MHz channel and broadcasted to all users.

A cable head end combines the downstream data channels with other services such as video, pay-per-view, audio, or local station programs that are typically received by television subscribers over the TV cable network. The combined signal is then transmitted throughout the cable distribution network and at the subscriber location. The television signal is received by something such as a set top box, whereas user data is separately received by a cable modem box and sent to a PC. Figure 9.9 illustrates this procedure.

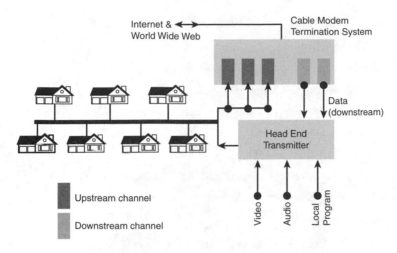

Internet & World Wide Web

Cable Modem Termination System

Data (downstream)

Head End Transmitter

Video · Audio · Local Program

Upstream channel

Downstream channel

FIGURE 9.9

Cable modem termination system and cable head end transmission.

Similar to DSL systems, a cable data system is comprised of many different technologies and standards. To develop a mass market for cable modems, products from different vendors must be interoperable. To accomplish the task of interoperable systems, the North American cable television operators formed a limited partnership— –Multimedia Cable Network System (MCNS)— and developed an initial set of cable modem requirements. This document set is called DOC-SIS. Comcast, Cox, TCI, Time Warner, Continental (now MediaOne), Rogers Cable, and CableLabs initially formed the MCNS partnership. The DOCSIS requirements are now managed by CableLabs, which also administers a certification program for vendor equipment compliance and interoperability.

At the cable modem physical layer, the downstream data channel is based on North American digital video specifications (that is, International Telecommunications Union [ITU]–T Recommendation J.83 Annex B), which includes the following features:

- 64 and 256 QAM
- 6 MHz–occupied spectrum that coexists with other signals in cable plant
- Concatenation of Reed-Solomon block code and Trellis code that supports operation in a higher percentage of the North American cable plants
- Variable length interleaving support for both latency-sensitive and latency-insensitive data services
- Contiguous serial bit-stream with no implied framing, which provides complete physical (PHY) and MAC layer decoupling

The upstream data channel is a shared channel with the following features:

- QPSK and 16 QAM formats
- Multiple symbol rates
- Data rate support from 320 Kbps–10 Mbps
- Flexible and programmable cable modem under control of CMTS
- Frequency agility
- Time-division multiple access (TDMA)
- Support of both fixed-frame and variable-length protocol data units
- Programmable Reed-Solomon block coding
- Programmable preambles

Privacy is a big issue with cable modem networks because cable modems are part of a broadcast network. Privacy of user data is achieved by encrypting link-layer data between cable modems and CMTS. For this purpose, the cable modem and CMTS head end controller encrypt the payload data of link-layer frames transmitted on the cable network. The Security Association (SA) assigns a set of security parameters, including keying data, to a cable modem. All of the upstream transmissions from a cable modem travel across a single upstream data channel and are received by the CMTS. In the downstream data channel, a CMTS must select appropriate SA based on the destination address of the target cable modem. Baseline privacy employs the data encryption standard (DES) block cipher for encryption of user data. The encryption can be integrated directly within the MAC hardware and software interface.

Power Lines Communication (PLC)

At first glance, provisioning network access the electric power line infrastructure seems to be the most straightforward approach that provides the largest possible coverage. That is because close to 100% of all potential buildings of interest are connected to the power grid. Power outlets are located in virtually every room within a building. However, the copper-based wire infrastructure was never intended to be used for high-speed data network access. For all practical purposes, only lower frequency (bandwidth) signals can travel over power lines.

How PowerLine Works

Northern Telecom and United Utilities originally developed Digital PowerLine technology. Electrical utilities can transmit regular low-frequency signals at 50–60 Hz and much higher frequency signals above 1 MHz without affecting either signal. The lower frequency signals carry power, whereas the higher frequency signals can transmit data. The technology is capable of transmitting data at a rate of 1 Mbps over existing electricity infrastructure. However, sophisticated digital processing and "conditioning" of the existing electricity infrastructure are required to achieve these kinds of data rates.

Digital PowerLine uses a network, known as a High Frequency Conditioned Power Network (HFCPN), to transmit data and electrical signals. An HFCPN uses a series of conditioning units (CUs) to filter those separate signals. The CU sends electricity to the outlets in the home and data signals to a communication module or "service unit." The service unit provides multiple channels for data, voice, and so on. Base station servers at local electricity substations connect to the Internet via fiber or broadband coaxial cable. The end result is similar to a neighborhood local area network.

PowerLine Equipment

The Digital PowerLine base station is a standard rack-mountable system designed specifically for current street electricity cabinets. Typically, one street cabinet contains 12 base station units, each capable of communicating over 1 of 40 possible radio channels. These units connect to the public telecommunications network at E1 or T1 speeds over some broadband service.

Several options, with different costs, can provide broadband Internet service to each base station. The simplest solution is connecting leased lines to each substation. This solution is potentially quite costly because of the number of lines involved. A wireless system has also been suggested to connect base stations to the Internet. This option reduces local loop fees, but increases hardware costs. Another alternative involves running high-bandwidth lines alongside electric lines to substations. These lines could be fiber, ATM, or broadband coaxial cable. This option avoids local loop fees, but is beset by equipment fees. The actual deployment of Digital PowerLine will probably involve a mix of these alternatives, optimized for cost efficiency in different areas and with different service providers.

The base stations typically serve approximately 50 customers, providing more than 20 MHz of usable spectrum to near-end customers and between 6–10 MHz of useable spectrum to far-end customers. The server operates via IP to create a LAN-type environment for each local service area.

The CU for the Digital PowerLine Network is placed near the electric meter at each customer's home. The CU uses band pass filters to segregate the electricity and data signals, which facilitate the link between a customer's premise and an electricity substation.

The CU contains three coupling ports, as shown in Figure 9.10. The device receives aggregate input from its Network Port (NP). This aggregate input passes through a high-pass filter. Filtering allows data signals to pass to a Communications Distribution Port (CDP), and a low-pass filter sends electric signals to the Electricity Distribution Port (EDP).

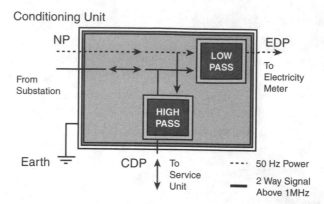

Figure 9.10

The conditioning unit for a Digital PowerLine Network.

The 50 Hz signal flows from the low-pass filter, out of the EDP, and to the electricity meter. The low-pass filter also serves to attenuate extraneous noise generated by electrical appliances at the customer premises. Left unconditioned, the aggregation of this extraneous noise from multiple homes would cause significant distortion in the network.

The high-pass filter facilitates two-way data traffic to and from the customer premise. Data signals flow through the CDP to the customer's service unit via standard coaxial cable.

The service unit is a wall- or table-mountable multipurpose data communications box. The unit facilitates data connections via BNC connectors to cable modems and telephone connections via standard line termination jacks. Alternative Differential Pulse Code Modulation (ADPCM) is used for speech sampling. Because Digital PowerLine allows for the termination of multiple radio signals at the customer premises, the service unit can facilitate various Customer Premises Equipment (CPE) simultaneously. In a manner similar to ISDN, data (computers) and voice (telephones) devices can coexist without interfering with each other.

PowerLine Deployment Strategy

In the Digital PowerLine model, small LANs are created; they terminate at each local electricity substation. These LANs will share a T1/E1 connection to the Internet, similar to a corporation leasing a T1 line. Individual users should experience tremendous speed increases over conventional 28.8 Kbps or 56 Kbps dial-up connections, even at peak usage. Only the substation server equipment and customer conditioning/service units need to be installed to establish a Digital PowerLine network.

Dedicated, multipurpose communication lines make the Digital PowerLine model an attractive option for the information age. Wide bandwidth and frequency division multiplexing allow for multiple lines to a single household. Ideally, an entire family could utilize their own communication devices simultaneously, whether telephone or PC, without interrupting one another.

The Digital PowerLine model has many possible extensions. Those mentioned in reviews and technical journals include "the wired home" and remote customer information services.

PowerLine Uses

Because Digital PowerLine creates a LAN-type environment by running IP, people could theoretically control all of the appliances in their home from their PC or a remote device. Each home on the neighborhood LAN would operate as a subnetwork of the LAN, and each electrical outlet could be treated as a node on that subnetwork.

Remote services such as remote metering have already been tested under this model, and many more services are possible. Because the service provider can keep track of electricity and bandwidth usage via the network, customers will also be able to monitor their usage, reliably predict billing, and keep an eye on household usage.

PowerLine Issues

Several implementation issues have held back Digital PowerLine in North America and the UK. Respectively, the problems are the numbers of users per transformer and the size and shape of light poles.

In North America, a transformer serves from 5–10 households, whereas in Europe, a transformer serves 150 households. Digital PowerLine signals cannot pass through a transformer. Therefore, all electrical substation equipment needed for Digital PowerLine has to be located after the transformer. Because there are fewer households per transformer in North America, predicted equipment costs are prohibitive. However, this conclusion has been debated. Analysts suggest that 100% subscription rates are possible in the U.S., and that at such a rate, Digital PowerLine is profitable.

Soon after the first trials of Digital PowerLine in the UK, some unanticipated problems arose. Certain radio frequencies were suddenly deluged with traffic, making it impossible to transmit on those frequencies. BBC, amateur radio, and the UK's emergency broadcasting service were affected. The apparent culprits were standard light poles. Then it became clear that by pure chance, British light poles were the perfect size and shape to broadcast Digital PowerLine signals. This situation posed problems not just because of the frequencies involved but also because anyone could listen in on the traffic. The privacy issue has not been fully addressed at this point, besides suggestions that all sensitive information should be encrypted.

Among the three wire-based access technologies discussed in this chapter (phone lines, coax cable, and power lines), the power line system is certainly the least developed and commercially deployed approach. When compared to xDSL and cable modem systems, power lines are certainly the slowest-speed access alternative. Therefore, this technology won't be adapted in business environments that require higher speed networking access. For residential customers seeking lower bandwidth, the basic connectivity of the digital power line approach might be an inexpensive option.

In-home networking using existing in-wall power outlets recently gained more market acceptance. The equipment uses the HomePlug 1.0 standard for using power lines to connect devices within the home. It has been plagued with technical difficulties such as interference caused by use of electrical appliances, but some equipment manufacturers claim those problems have been solved. Power line networking opens a number of interesting possibilities because it enables network connections simply by plugging a HomePlug-enabled appliance into a wall socket.

More detailed information about power line networks can be found on the HomePlug Alliance Web page at `http://www.homeplug.org/index.html`.

LMDS

Local Multipoint Distribution System (LMDS) is a third-generation point-to-multipoint (PMP) radio system that provides access to broadband services, including highly asymmetrical or burst-type traffic. LMDS refers to a block of FCC licenses in the microwave region that is designed for broadband communications. All the decisions involving bandwidth usage—from modulation scheme to protocol—are left to the licensee.

LMDS Implementation and Uses

LMDS can be implemented in a variety of ways. Many features are common to all LMDS systems because of the way signals in that frequency band behave as they go through the atmosphere. LMDS can provide a distribution of services through a variety of multiple access techniques, such as Frequency Division Multiple Access (FDMA), Time Division Multiple Access (TDMA), and Adaptive Time Division Duplexing (ATDD). New transport platforms are also on the horizon, such as Wireless Asynchronous Transfer Mode (W-ATM) and Wireless IP, which will have the capability to consolidate services on an end-to-end network basis. W-ATM technology is capable of offering scalable bandwidth in combination with Bandwidth on Demand (BOD).

The licensed LMDS frequency bands in the U.S. are around 28–31 GHz. Other countries allow LMDS operators to use different frequency bands. In these higher microwave frequency bands, coverage is a critical issue because signals are easily obstructed. Although deployment of higher frequency, fixed, wireless point-to-point radios—such as 38 GHz radios—has proven that reliable links can be achieved at millimeter wavelengths, significant differences are associated with an LMDS multipoint access system. Point-to-point systems have the advantage of two fixed end points, allowing for confirmation of line of sight. In the multipoint system, only the hub location can be strategically selected. The subscribers are in fixed locations, and generally the customer premises' antenna units are mounted at or near the roofline as opposed to being mounted on towers.

Because LMDS operates in the higher microwave spectrum, the system requires line of sight between antenna locations that are typically mounted on building rooftops or tower structures. The primary benefit associated with line-of-sight operation is that, when coupled with directional antennas and varying polarization, the service area can be highly sectorized, allowing for reuse of the valuable spectrum resource within a Basic Trading Area (BTA). However, LMDS antennas, like all directional antennas, generate side lobes or concentrations of power in unintended directions. In the case of closely packed antennas (such as in an LMDS hub), these lobes need to be taken into consideration to ensure that no interference exists between antennas. Polarized antennas can help minimize this interference, although they do complicate the installation. The typical diameter of a cell is about one mile.

LMDS License Requirements

A license is absolutely necessary to operate an LMDS frequency because it is in an allocated spectrum. Licensing fees could be a significant portion of the expense of an LMDS system, but the fact that LMDS is allocated can help ensure that there will not be unexpected interference from other systems potentially operating in the same spectrum. The FCC attempts to minimize the possibility of interference by planned logistics, and in the event of difficulties, assists in mitigation by becoming the arbitrator.

Another advantage to working in the LMDS frequency band is that the spectral bandwidth in the higher frequency band is typically larger, allowing for higher bandwidth to be transported over the system. In addition, high gain antennas are relatively small; an antenna providing a beam width of less than 3 degrees and a gain of 35 dBi can be designed to be less than 12 inches in diameter.

LMDS Path Loss Issues

Significant challenges are associated with deploying LMDS systems at 28–31 GHz. In addition to excessive path loss associated with this band of frequency, rainfall attenuation must also be taken into account.

The theoretical cell size will largely be governed by the free-space path loss of radio frequency waves. Such loss is typically the largest single component of attenuation in the system. This attenuation is proportional to the following:

$$\left(4\pi\,\frac{r}{\lambda}\right)^2$$

where λ is the wavelength of the carrier frequency and r is the radius of the cell size. Given that the wavelength is fixed based on the LMDS spectrum that is purchased, the cell size is the only controlling factor for the largest system loss component. This will be a significant factor in determining the amount of margin available in the system to deal with other environmental factors, such as rainfall and foliage.

Rainfall is the most significant path degradation for LMDS systems. Because the wavelength of the signals involved (~10 mm at 28–31 GHz) is about the size of raindrops, scattering and attenuation can result. In addition, rainfall causes depolarization of the LMDS signals, decreasing the desired signal level and leading to poor interference isolation between adjacent sectors and cell sites. Attenuation figures will vary by regional deployment and will impact cell site planning by reducing acceptable cell sizes.

Because rainfall varies considerably from one part of the country to another, an effective deployment in Los Angeles or Phoenix might not work well in Miami or Houston. Service providers address these problems in two ways to mitigate potential availability problems caused by rainfall attenuations. One way is to simply account for expected rainfall in a link budget. The network design engineer takes historical rainfall patterns into account and adjusts the node (antenna) placement and power levels to account for worst-case rainfall. High rainfall rates typically result in smaller diameters of cells. This increases the deployment cost for the service provider because more antennas are needed to cover a specific area. On the other hand, smaller cells also imply fewer subscribers per cell, which results in a higher data rate per subscriber. A higher data rate means more or higher bandwidth services can be provided over the network, which generates more revenues for the service provider.

The LMDS network topology envisioned by most LMDS vendors is a fiber-fed, hub-and-spoke network. This network consists of a series of LMDS hubs that contain a set of antennas to feed a variety of customers. These hubs are in turn connected with fixed fiber connections. However, despite the potential cost advantage of the anticipated point-to-multipoint topology, the majority of current LMDS deployments are point-to-point systems.

LMDS Capacity

The system capacity (aggregate data rate available) is a function of many variables. They can be electrical (bandwidth, modulation scheme, frequency reuse) and environmental (path length or cell size, line-of-sight coverage, rainfall) in nature. In the United States, the largest contiguous block of spectrum for LMDS is 850 MHz (27.5–28.35 GHz). The data rate is dependent on the modulation scheme used. 64-QAM, for example, can provide a total of up to 4,250 Mbps total network capacity, which would then be divided into upstream and downstream traffic. The LMDS backbone can be partitioned symmetrically or asymmetrically with respect to data rate. Thus, part or none of the total capacity can be allocated for two-way communication.

In reality, the total capacity is much lower. A 64-QAM modulation scheme requires an extremely high signal-to-noise ratio (SNR) at the receiver to reach a carrier class bit-error rate (BER), which is typically in the range of 10^{-12} to 10^{-14}. This severely limits cell sizes and introduces greater receiver costs to meet the required SNR. More practically, a 16-QAM system with forward-error correction (FEC) can yield approximately 2,700 Mbps total bandwidth.

The capacity present within the LMDS backbone can be partitioned among many users in a nearly arbitrary fashion. It is important to note, however, that the addition of users introduces a system overhead that will consume additional data-rate capacity. Because several users must all have access to the channel, an appropriate multiple-access scheme must be used. Typically, this is either done in the frequency division duplex (FDD) or time division duplex (TDD) domain.

Ultimately, the end user will be interested in the total capacity available to the individual, typically in a bidirectional configuration for residential or multitenant commercial buildings. Given a symmetrical, bidirectional configuration, the 2,700 Mbps can be reasonably partitioned to, at most, 25–30 DS-3 (44.7 Mbps) class customers. Most of the currently deployed LMDS systems do not operate at these high speeds. T1/E1 speeds are more common than higher speed data rates. A DS-3 data rate is likely to be sufficient for residential customers and small- to medium-size businesses in the short and intermediate term; however, large businesses and multitenant units can be expected to outgrow this bandwidth within a few years.

LMDS Coverage Area

The total coverage area of an LMDS is dictated by the license, which sets forth geographic boundaries of the BTA. Within that, the area can be subdivided into cells based on the following criteria:

- Line of sight: Because LMDS is a line-of-sight technology, the receiving antenna must be able to "see" the transmitting antenna.
- Data-rate and total user trade-off: Typically, a larger cell will enable more users to be serviced. However, the path loss of the RF signal will grow by the square of the cell radius. This will reduce the amount of the available link margin. For a given bit-error rate requirement, a way to regain that margin is to reduce the signal-to-noise ratio that is required at the receiver. Typically, this is accomplished by changing modulation schemes

9

ALTERNATIVE
ACCESS
TECHNOLOGIES

and using less dense modulation schemes (constellations). One example would be to change the modulation from a quadrature amplitude modulation scheme, such as 16-QAM, to a quadrature phase shift keying (QPSK). However, this directly reduces the amount of bandwidth that is transported within the cell. Therefore, the service provider must perform a market analysis to project the user base and its bandwidth needs to be able to effectively design cell sites and sizes.

LMDS Deployment

The speed with which a system can be deployed and installed depends on several factors. Assuming that spectrum acquisition is complete, several planning steps must be undertaken before a system can be deployed. These specific steps typically include the following:

1. Terrain and building mapping to determine line-of-sight coverage, RF shadowing, side-lobe interference, and cell boundaries. Antenna positioning will depend on coverage, projected demand, and cell sizes.

2. In conjunction with step 1, whether data must be incorporated into the mapping to determine rainfall statistics. Such statistics will assist in determining attenuation ("rain fade"), which might require reduction in cell sizes or reduction in bit rate for a given cell size.

3. Acquisition of roof rights for the buildings in question.

4. Installation, antenna pointing, tuning, cabling, and testing. In conjunction with this, signal strength should be measured at all sites under a variety of conditions to ensure reliability and that signal loss or shadowing is not occurring.

Like most complex network systems, LMDS installations can be expensive, and the service provider needs to cover a variety of upfront costs before the actual deployment can take place. These costs include spectrum licensing fees, roof rights access and fees, network equipment, customer premise equipment, and management equipment.

MMDS

In 1970, the FCC created the multipoint distribution system, which was called MDS at that point in time. MDS was created to give licensed operators the opportunity to use MDS channels to transmit digital data or television programs as a business service within a 30-mile radius of a community. The initial application scenario was certainly heavily focused on TV broadcast, and the aspect of digital data transmission was a minor focus. The MDS system utilized the microwave spectrum in the 2.1–2.7 GHz range, while being able to provide multipoint network services.

The first efforts turned out to be a disaster. High equipment costs and unreliable technology took the MDS idea out of the market for more than a decade. In the early 1980s, recognizing the cost differences between satellite dish receiving stations and an MDS transmission system, pay-per-view programmers started to utilize the MDS system again to transmit to signal-to-cable head ends, hotels, and condominiums. MDS systems also supported another group of 30 NTSC-formatted channels, known as instructional television fixed service (ITFS). Schools and universities typically use the ITFS channels to deliver instructional TV courses to classrooms and campuses throughout a specific region.

Today's multichannel MDS (now called MMDS for multichannel multipoint distribution system) reflects the many recent changes brought by the 1996 Telecommunications Act. At that point in time, the FCC converted eight ITFS channels to full-time MMDS channels to encourage competition with local cable TV services. To further increase the channel capacity, the FCC authorized the ITFS channel operators to lease out air time to local MMDS operators for TV programming services at night when they were not being used. The combination of full-time and part-time channels makes up the 31 channels of the MMDS system today. Before getting into the data-delivery aspect of MMDS systems, this chapter will briefly review the technology from the TV channel operator's point of view because this was the original focus of MMDS systems.

MMDS Transmission Technology

The MMDS transmission system utilizes frequencies in the microwave spectrum. Although not as strict as a requirement for higher frequency microwaves, a clear line of sight between the transmitting antenna and the subscriber receiving antennas is beneficial. Trees and heavy foliage can severely attenuate or entirely block signals. The block diagram in Figure 9.11 shows an MMDS transmitter. The TV modulator used in the transmitter is the same as the modulator used in the cable TV (CATV) system.

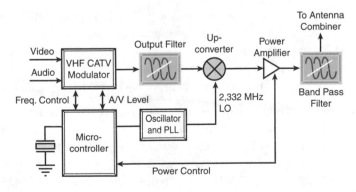

FIGURE 9.11
MMDS transmitter circuitry.

The up-converter converts the output channel frequency to the desired MMDS channel frequency. This signal is amplified and fed to the antenna system after appropriate band pass filters filter it. The up-converter local oscillator (LO) is a highly stable microwave frequency source. The LO frequency is 2,332 MHz, with a stability better than +/-0.0002 PPM.

Recent advances in microwave ICs have led to high integration of many of their complex digital-circuit functions, such as a microwave prescaler, PLL frequency control, and digital video synchronizing. Components that offer more capabilities and greater overall system performance have dramatically changed the RF circuit design and PC layout. In addition, these products offer high noise immunity and improved phase stability. Until recently, microwave ICs have primarily been used in sophisticated custom and military applications. As a consequence, the majority of MMDS transmitters operating today are using the frequency "offset" scheme to combat cochannel interference. Similarly, new wireless technologies developed for cordless phones and data communication products can be designed into many circuit functions in the transmitter design. The signal is being generated in the digital domain, which allows precise control of frequency and phase with a suitable digital-signal control system.

In preparing for an FCC license application, the system operator is required to submit the same documentation to the FCC with the license application:

- Interference analysis for the proposed area and coverage maps that indicate the signal levels
- Antenna tower height, radiation patterns, and terrain profiles of the covered area
- Installation instructions for all operating equipment
- Test and alignment procedures
- Equipment specifications

Along with equipment procurement and installation, the early stage of engineering study might involve completing most of the FCC-regulation compliance work. The growth of wireless services in general has led many communities to stop construction of antenna sites. System engineering design and planning must be well aware of this trend. This can lead to long and daunting regulatory obligations that delay the deployment of MMDS systems.

Figure 9.12 shows the block diagram of the low-noise block (LNB) downconverter that is used at the receiver side. The key parameters are antenna impedance match, noise figure, conversion gain, overall linearity, and LO stability. The converter LO consists of a voltage-controlled oscillator (VCO) that is phase-locked to a reference frequency by a fixed crystal oscillator. The LO frequency value is fixed at 2,278 MHz because the input frequencies of MMDS channels are 2,500 MHz to 2,686 MHz. Consequently, the block downconverter output signal frequencies range from 222–406 MHz. These signals are fed to the TV set or set top box input. The integration of the downconverter and overall received TV signal will be determined by transmitter power, antenna height, gain, cross-polarization, and front-end noise figure. For an NTSC video signal, a signal-to-noise ratio (SNR) of 47 dB will produce good to excellent TV reception. A value of less than 35 dB produces poor picture quality.

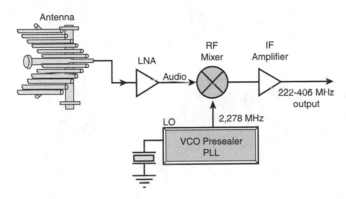

FIGURE 9.12
Block diagram of the downconverter at the customer premise.

MMDS Systems for Data Applications

MMDS service providers look at MMDS service as an untapped revenue stream for Internet service. For instance, these providers envision large office buildings, hotels, and motels to be served by the system's point-to-multipoint network topology.

The value proposition is that any system with a single omnidirectional antenna can service an area 20 miles in diameter with 30 TV channels and two 30 MB downstream data channels. This could potentially translate into as many as thousands of modem connections. However, for the wireless ISP operators to provide an Internet-based network, a robust, reliable, and cost-effective interactive system is required. The ISPs will have to determine the best "multicast protocol," one that offers the fastest and most information-rich environment to encourage subscribers to carry e-mail and Internet data applications linked with the carrier's broadband network infrastructure. This will ultimately determine the success of service that carriers provide.

Unlicensed Microwave Systems

Unlicensed radio frequency (RF)/microwave systems have been widely deployed over the past few years. If it weren't for complaints of interference problems, they would "own" the fixed wireless marketplace. Where fog is the enemy of FSO, interference is the enemy of RF.

ISM Band Operation (Spread Spectrum Technology)

In 1985, the FCC (Federal Communications Commission) allocated three frequency bands for a radio transmission technique known as spread spectrum communications. These frequency bands are known as Industrial Scientific Medical (ISM) bands. The spread spectrum transmission technology was originally developed by the military. Spread spectrum transmission has much greater immunity to interference and noise compared to conventional radio transmission. In addition, an increasing number of users can use the same frequency, a feature similar to cellular radio systems.

Methods of Signal Spreading

Under the regulations, users of FCC-certified spread spectrum products do not require a license from the FCC. The only requirement is that the manufacturers of spread spectrum products must meet FCC spread spectrum regulations, one of the main regulations being related to the maximum amount of launch power. Spread spectrum is a technique that takes a narrow band signal and spreads it over a broader portion of the radio frequency band. This has the operational advantage of being resistant or less susceptible to electromagnetic interference. However, due to unfounded concerns over the increased frequency space it occupies, the FCC, until recently, did not permit commercial use of the technology.

The FCC rule changes in 1985, combined with the continuing evolution of digital technology and the demand for access bandwidth, has catalyzed the development of spread spectrum data communication radios. In performing spread spectrum, the transmitter takes the input data and spreads it in a predefined method. Each receiver must understand this predefined method and despread the signal before the data can be interpreted at the subscriber location.

Frequency Hopping and Direct Sequencing are the two basic methods to perform the signal spreading. Frequency hopping spread spectrum (FHSS) spreads its signals by "hopping" the narrow band signal as a function of time. Actress Hedy Lamarr and composer George Antheil originally conceived the idea of FHSS during World War II. FHSS employs a narrowband carrier, changing its frequency in a pattern known only to the sender and receiver. As intended, this makes information difficult to intercept. Direct sequencing spread spectrum (DSSS) spreads its signal by expanding the signal over a broad portion of the radio band.

Spread Spectrum Frequencies

The FCC allows the use of spread spectrum technology in three radio bands—902–928 MHz, 2,400–2,483.5 MHz, and 5,752.5–5,850 MHz—for transmission under 1 Watt of power. This power somewhat limits the interference within the band over long distances. Table 9.2 shows the allocation and spectral bandwidth of unlicensed microwave bands. Also shown is the spectrum location of the "Unlicensed National Information Infrastructure" (U-NII) band. The 300 MHz spectrum is divided over three 100 MHz bands between 5–6 GHz. The upper U-NII and the 5.8 GHz ISM band overlap in the frequency spectrum.

TABLE 9.2 Allocation of Unlicensed Transmission Bands in the Microwave Spectrum

Block	Frequency	Bandwidth	Current Uses
ISM	900 MHz	26 MHz	Cordless phones; remotes
ISM	2.4 GHz	100 MHz	Cordless phones, microwave ovens, wireless LANS, backhaul, local access
ISM	5.8 GHz	150 MHz	Wireless LANs, backhaul
U-NII	5 GHz	300 MHz	Local access, backhaul

Spread Spectrum Advantages

Some advantages of spread spectrum systems operating in the ISM bands include the following:

- No FCC site license: The FCC will grant a one-time license on the radio product. After that license is granted, the product can be sold anywhere in the U.S. Some countries outside the U.S. also allow unlicensed spectrum operation in the 900 MHz and 2.4 GHz bands. However, most of those countries are more stringent with respect to the launch power than the U.S. is.

- Low rain fade: Unlike other microwave solutions operating in outdoor environments at frequencies above 10 GHz, the lower microwave bands are much less impacted by rain.

- Interference immunity: Spread spectrum radios are inherently more immune to noise than conventional radios. Spread spectrum radios will operate with higher efficiency than conventional technology. In densely populated areas such as metropolitan centers, interference can be a major problem.

- Multichannel: Conventional radios operate on a specific frequency controlled by a matched crystal oscillator. The specific frequency is allocated as a part of the FCC site license, and the equipment must remain on that frequency. Low power devices such as cordless phones are an exception. Spread spectrum data radios offer the opportunity to have multiple channels, which can be dynamically changed through software. This allows for many applications, such as repeaters, redundant base stations, and overlapping antenna cells.

Spread Spectrum Standards

One widespread standard today dominates much of the commercial ISM band spread spectrum market: 2.4 GHz. This standard is the IEEE 802.11 standard. The IEEE 802.11 specification is a wireless LAN standard developed by the Institute of Electrical and Electronic Engineering (IEEE) committee to specify an "over the air" interface between a wireless client and a base station or Access Point, as well as among wireless clients. First conceived in 1990, the standard has evolved from various draft versions (Drafts 1 through 6), with approval of the final draft on June 26, 1997.

Like the IEEE 802.3 Ethernet and 802.5 Token Ring standards, the IEEE 802.11 specification addresses both the Physical (PHY) and Media Access Control (MAC) layers. At the PHY layer, IEEE 802.11 defines three physical characteristics for wireless local area networks: diffused infrared, direct sequence spread spectrum (DSSS), and frequency hopping spread spectrum (FHSS).

Whereas the infrared PHY operates at the baseband, the other two radio-based PHYs operate at the 2.4 GHz band. For wireless devices to be interoperable, they have to conform to the same PHY standard. All three PHYs specify support for 1 Mbps and 2 Mbps data rates. Although FHSS lives on in some products, it is not part of 802.11b. DSSS was the encoding scheme that became part of the 802.11 standard. This type of signaling uses a broadband carrier, generating a redundant bit pattern (called a "chip") for every bit of data to be transmitted. Although DSSS is obviously more wasteful in bandwidth, it copes well with weak signals. Data can often be extracted from a background of interference and noise without having to be retransmitted, making actual throughput superior. DSSS provides a superior range, and is more capable of rejecting multipath and other forms of interference. The 802.11b version of DSSS transmits data at a nominal 11 Mbps. (Actual rates vary according to distance from another transmitter/ receiver.) It is downwardly compatible with 1 Mbps and 2 Mbps wireless networking products, provided they also use DSSS and are 802.11-compatible.

With few exceptions, 802.11b is a worldwide standard in the 2.4–2.48 GHz frequency band, dividing this into as many as 14 different channels. In the U.S., 11 channels are available for use.

Spread Spectrum Market

The target market for 802.11 products was mainly directed toward indoor wireless LAN applications. Inside buildings, a 2.4 GHz signal can penetrate through walls. The typical coverage area of a wireless access point can extend several hundred feet depending on the properties of the walls. Vendors must tailor their hardware access points to use legal channels in each country where they ship. Wireless NICs, however, can often adapt automatically to whatever channels are being employed locally. Therefore, it is possible to travel with an 802.11b client and make connections to 802.11 equipment in another country.

Driven by the need of inexpensive access solutions, outdoor applications using the unlicensed ISM band in conjunction with the 802.11 LAN standard became popular among corporate LAN users. The typical task was to interconnect buildings in a campus-like LAN environment.

Several buildings could be interconnected by using either higher power grids or Yagi antennas in point-to-point scenarios, or by using a central omnidirectional antenna at one central location. However, for outdoor applications over longer distances, line of sight is a requirement. Therefore, for a point-to-multipoint network, the omnidirectional antenna has to be placed at a location that can be seen from all other remote networking locations.

This installation strategy is similar to the installation of LMDS systems. The typical coverage area of such an ISM band network could extend over a few miles in diameter. However, in densely populated environments such as metropolitan areas, the unlicensed ISM band approach was never adapted on a larger scale. The main reason for this is related to the unlicensed nature of this technology: When more organizations started to use this band, the noise floor increased and they made the ISM band unusable. Even the spread spectrum approach was no longer efficient enough to prevent signal interference. Therefore, commercial carriers providing high-quality network access never seriously considered the ISM band as an alternative access technology. However, some ISPs that operate in less populated or rural areas use this technology to provide commercial network access. Some local communities also use the 2.4 GHz ISM band to build local access data networks. More information regarding the 802.11b community network can be found at http://www.toaster.net/wireless/aplist.php.

Faster flavors of business-class wireless LANs are under development, and several versions, such as 802.11a, 802.11g, and HiperLAN/2, are contending for status as the next 802.11x standard. Figure 9.13 shows the typical access topology of an 802.11 wireless network.

U-NII Band Systems

The European Community (ETSI) was the first to open the 5 GHz band. So far, the 5.2 GHz band is dedicated to a standard called High Performance Local Area Network, or HIPERLAN, and the 5.4 GHz band is reserved for HIPERLAN II. As ETSI has done for GSM, only systems that fully conform to those standards on the Physical layer and MAC layer can operate in the band.

U.S. Adoption of U-NII at 5.x GHz

Following the European effort, in January 1997, the FCC made 300 MHz of spectrum for Unlicensed National Information Infrastructure (U-NII) available for deployment of unlicensed radio systems. The rules were liberal. To limit systems, FCC incorporated complicated power rules that made the use of roughly 20 MHz bandwidth optimal. As a general rule, systems using less bandwidth could transmit less power, and systems using more bandwidth would not get more power. The FCC made this move in the belief that the creation of the U-NII band (besides the already established unlicensed ISM band) would stimulate the development of new, unlicensed digital products to provide efficient and less expensive solutions for local access applications.

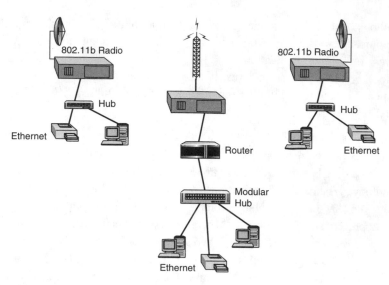

FIGURE 9.13

Example of an 802.11 wireless ISM band network topology.

U-NII Band Allocation

The U-NII band is divided into three 100 MHz subbands around 5.2, 5.3, and 5.8 GHz. The first band is strictly allocated for indoor use and is consistent with frequency allocation of the European High Performance Local Area Network (HIPERLAN). The second and third bands are intended for high-speed, digital, local access products for "campus" and "short-haul" microwave applications. Table 9.3 shows the FCC allocation of the U-NII band spectrum and the corresponding power levels allowed in each of the three bands.

TABLE 9.3   The FCC U-NII Band Standard

	Band 1	*Band 2*	*Band 3*
Frequency	5.15–5.25 GHz	5.25–5.35 GHz	5.725 to 5.825 GHz
Power (max)	200 mWatts (EIRP)	1 Watt (EIRP)	4 Watts (EIRP)
Intended Use	Indoor use only	Campus applications	Local access, 10 miles

The FCC rules for products reveal that operation in bands 2 and 3 of the U-NII spectrum is best suited for digital microwave applications over distances of approximately 10 miles. The assigned 100 MHz spectrum in each of these bands, in combination with the maximum power levels allowed, facilitate the deployment of medium capacity and reliable microwave links for both data and telephony transmission. Figure 9.14 shows the relationship between the maximum EIRP (Equivalent Isotropic Radiated Power) and the occupied bandwidth of the transmitted signal in accordance with the regulations.

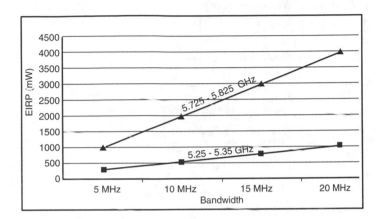

FIGURE 9.14

Maximum EIRP in the FCC U-NII bands 2 and 3 (Courtesy of Wireless Inc.).

To ensure the most effective use of the band, vendors decided to use robust modulation schemes, such as Binary Phase Shift Keying (BPSK), Frequency Shift Keying (FSK), or Quadrature Phase Shift Keying (QPSK). These schemes are capable of high-speed Ethernet or multiple T1/E1 digital circuits and are cost effective.

Similar to the ISM band and unlike high frequency microwave links above 10 GHz, the U-NII band is not affected by outages due to rain attenuation. Microwave transmission is also less affected by free-space loss at 5.25 < 5.825 GHz than high frequency microwave. Some U-NII band equipment manufacturers claim that even with FCC limitations on power output and antenna gain in the U-NII bands, a system operating at 5.3 and 5.7 GHz over 10 miles can archive 99.995% availability. By using both bands 2 and 3, microwave paths can operate in full duplex mode, meaning information can be transmitted and received simultaneously.

The microwave system performance using both the 5.3 GHz and 5.7 GHz bands is limited by the FCC transmitter and antenna rules for the second band. The use of dual band operation, however, does have the benefit of separating the system transmitters and receivers by approximately 480 MHz. This significantly simplifies the equipment transmitter and receiver design, resulting in a lower cost product. Dual band operation also promotes frequency reuse, allowing the use of 200 MHz of bandwidth as opposed to 100 MHz in single band operation.

Whereas some vendors implemented U-NII band systems capable of operating in a point-to-multipoint scenario, other vendors implemented point-to-point solutions. In the U-NII band, the use of 2- or 4-foot highly directional parabolic antennas (with gains of approximately 27 and 33 dBi, respectively) can improve the overall performance of the system. This benefits the vendors of longer distance point-to-point systems. However, the installation of larger antennas complicates the installation process. As shown in Figure 9.15, a high level of availability is typically achievable for path lengths below 10 miles. In parabolic antennas, the gain can exceed 6 dBi as long as the peak power spectral density is reduced proportionately. Parabolic antennas also offer additional isolation from co-located or adjacent microwave signals.

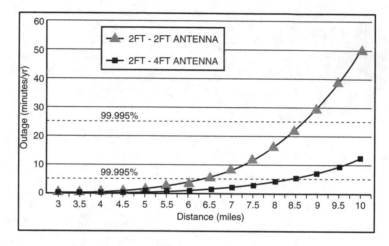

FIGURE 9.15

Typical microwave path performance in the U-NII band (courtesy of Wireless Inc.).

Some vendors have tried to come up with a stricter set of rules for the U-NII band, but they couldn't accommodate the conflicting requirement of all parties. A small group of network providers has tried with limited success to implement a U-NII band infrastructure. Most likely, private users currently in need of a cost-effective solution for short-haul access can find the greatest benefit of the U-NII band.

Because the U-NII is an "unlicensed" band, the costs and time associated with frequency coordination and licensing are eliminated. As with any other unlicensed microwave technology, network providers, businesses, schools, and government agencies can rapidly install microwave links for high-speed, digital local access. In the 5 GHz band, because of the availability of more bandwidth, higher speeds are possible (10–480 Mbps). However, operating in a higher frequency band increases the noise level, makes obstacles and walls more opaque to transmissions, and requires more SNR (Signal Noise Ratio). This means a reduced range compared to 2.4 GHZ products.

Fiber Access

The deployment of optical fiber in local loop access network has been extensively discussed for more than a decade. When compared to any other access media, optical fiber is certainly the ultimate media for high-speed access. Optical fiber access means ultra-high bandwidth, scalability, and reliability in one media. However, laying a new fiber infrastructure in the local loop also means ultra-high costs. In addition, it's more time consuming.

Just think about how long it took the phone companies to run telephone wires to each building to provide basic telephone service. In some rural area, that still didn't happen due to the high cost of deployment. As a matter of fact, the high cost and the deployment time of laying fiber were the main reason that technologies such as xDSL and cable modems that use the existing copper infrastructure were developed. Wireless service providers would not have been able to raise billions of dollars for microwave licenses if investors would have been convinced that fiber deployment in the local loop access market would be a viable short- or medium-term alternative. To understand the struggling past of fiber deployment in the local loop, it is helpful to briefly review the history of these efforts.

A Brief History of Network Access Deployment

The state of access networks has not changed significantly in decades for most end users who are seeking network access. Despite the heavy spending on network upgrades in recent years, the bulk of major U.S. carriers' investment has gone into increasing capacity in their core national networks to accommodate the growth of data and Internet traffic. With respect to optical fiber, most of the money spent for infrastructure deployment ended up in long-haul fiber deployment, and to some extent in metropolitan core networks. As a direct result of this spending activity, end users have experienced a significant decrease in the cost of long-haul traffic and long-distance voice traffic.

A larger interest in deploying a fiber-based local loop access infrastructure began at the beginning of the 1990s. Some Regional Bell Operating Companies (RBOCs), multiple cable system operators, and other service providers started to see the revenue opportunity in providing broadband access for residences and small business. The primary service motivating this interest was combining voice and video, which was generally known as Video Dial Tone. Many vendors developed systems to address this interest, including passive-optical-network (PON) systems, as well as digital-loop-carrier (DLC)-based fiber-to-the-curb (FTTC) systems. At this time, asymmetric DSL was in its early stages.

Over the next couple of years and by the mid-1990s, a large number of trials had been performed on all these products in an effort to evaluate both the technical feasibility as well as the possible service offerings. Even though many of the trials were considered successful, none resulted in

9

ALTERNATIVE
ACCESS
TECHNOLOGIES

large-scale deployments. Around that time, the interest of the RBOCs shifted from all-optical to a less aggressive Hybrid Fiber Coaxial (HFC) system approach. Another round of trials followed, which still did not result in significant deployments of HFC systems by the RBOCs.

The Impact of the Internet

About the same time the Internet started to become an important factor in society, the limits of dial-up connections were clear to everyone who logged on. By the last years of the decade, it was data, not video, that really became the driving force behind broadband network deployment. Cable systems operators and telephone companies were now in a race to roll out a high-speed data service, using their existing networks and not an entirely new architecture. Consequently, cable operators chose cable modems, and RBOCs selected the twisted-pair copper-based xDSL technology to deliver these services. Fiber deployments in the local loop were still far behind the bold predictions that were envisioned at the beginning of the 1990s.

It has taken a full 10 years for the industry to finally select a technology (xDSL and cable modems) and begin deployments on a broad scale.

Flavors of Fiber Access

Although many discussions have taken place about how deep fiber will migrate into the network, the closest that fiber ever penetrated to these locations in large-scale deployments was through DLC (Digital Loop Carrier) services for telephone or HFC (Hybrid Fiber Coax) services for cable TV.

Digital Loop Carrier (DLC) systems are based on an FTTC architecture. FTTC refers to any system that brings fiber within a few hundred feet of the subscriber. However, in terms of a DLC architecture, this usually refers to the practice of extending fiber from a DLC remote terminal closer to the subscriber.

In a typical installation, an optical fiber route connects the central office with a remote terminal that is placed to serve from a few hundred to 2,000 subscribers (see Figure 9.16). From the remote terminal, services can be delivered on copper, or they can be extended on fiber to an optical-network unit (ONU) serving a very small area, typically fewer than 100 subscribers. Because such FTTC systems originally evolved from DLC, they are efficient for delivering voice and narrowband data services. DLC is widely deployed by local-exchange carriers in North America, serving approximately 30% of access lines. That gives the FTTC systems a natural market base to be served. However, because DLC was originally designed as a narrowband voice service delivery system, many existing DLC systems are not easily upgraded to provide the additional bandwidth required for high-data rate broadband services.

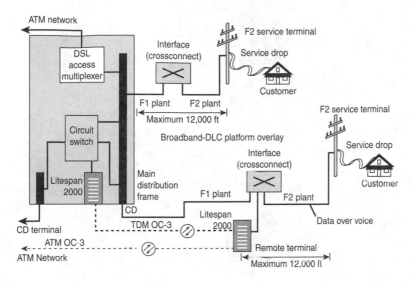

FIGURE 9.16

Digital loop carrier (DLC) network.

Hybrid Fiber Coax (HFC) is a derivation of the original all-coaxial cable TV network architecture that started appearing in North America about 50 years ago. Originally, these systems were strictly broadcast unidirectional video distribution networks. Because coaxial cables exhibit inherent high losses and the required amplifiers added noise and distortion to the system, the HFC cable infrastructure was far from perfect.

With the advent of low loss/high bandwidth optical fiber, cable system operators realized that they could improve the video performance of their networks. By incorporating optical fiber and broadband lasers into the coaxial cable plant, operators could decrease the length of the coaxial cable runs, reduce the amount of amplifiers needed, and reduce plant maintenance costs. Depending on the operator's design choices, a fiber node serves from 200 to about 2,000 subscribers (see Figure 9.17). During the 1990s, many cable systems were further upgraded to provide bidirectional transmission, and cable modems began to appear to allow subscribers to send data upstream over the HFC network.

Today, about 70% of all the cable plants in North America are capable of two-way traffic, and about 60% of residences are cable TV subscribers. Most cable system operators would like to be able to offer voice in addition to data service. However, as mentioned previously, the basic design of HFC limits the bandwidth that can be directed from subscribers into the network. Consequently, the technical challenges of providing data or voice service on this type of physical network are not trivial.

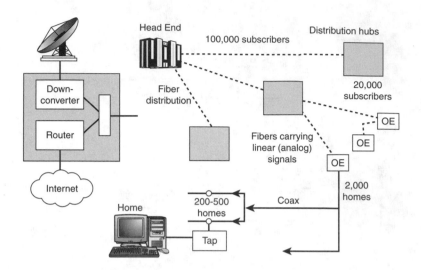

OE - Optical-electrical

FIGURE 9.17

Hybrid fiber coax (HFC) network.

However, neither the DLC nor the HFC systems extend the reach of the fiber close enough to the subscriber to overcome the limitations of the "last mile" technology. Although both DLC and HFC perform well at their basic mission, both systems had to undergo significant upgrades and improvements before they were capable of delivering limited bandwidth broadband data services.

PON Access Architectures

Most recently, the discussion around fiber in the local loop has been focused on two all-optical technologies, namely passive-optical-network (PON) architectures and point-to-point Gigabit Ethernet. Depending on the intended service, both techniques offer potential advantages to service providers and end users.

PON was originally developed and studied in the late 1980s. The key distinguishing feature of a PON architecture is the elimination of electronics from the last miles of the access network. This specific feature makes the PON architecture an all-optical architecture. Instead of using the remote terminal in a DLC-type system, passive optical splitting is employed to derive the individual fibers for individual end users. PON systems have the advantages of a relatively low cost for the distribution portion of the network. In addition, they provide savings in power, real estate, and maintenance costs associated with a remote terminal.

Although the passive splitting aspect of PON technology provides genuine cost advantages, it also presents some difficult technical challenges. Because it is a multiple-access system with variable ranges on all the branches, timing on PON systems must be extremely precise to ensure that end-user traffic is not lost or misdirected. In addition, capacity planning for PONs must be considered carefully when designing the physical network.

Second-generation PON systems have been under development for several years. Many startup equipment suppliers are planning to participate and compete in this market, along with several established access players. Major carriers, including several U.S. incumbent local-exchange carriers; European postal, telegraph, and telephone providers; and Asian service providers have been supporting this development activity through participation in the Full Service Access Network (FSAN) initiative (`http://www.fsanet.net`). FSAN defines a set of PON architectures using ATM as the transport technology (see Figure 9.18). FSAN is proposing PON as ITU draft standard G.983. Besides regarding ATM as a transport architecture platform, some PON vendors envision an Ethernet-based transport platform to be used in a PON architecture.

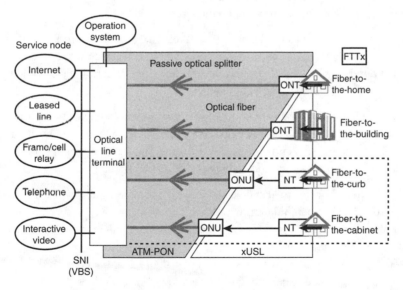

FSAN - Full Service Access Network
NT - Network terminal
ONT - Optical-network terminal
ONU - Optical-network unit
PON - Passive optical network

FIGURE 9.18

Full Service Access Network (FSAN) model for PON architectures using ATM as transport technology.

9

ALTERNATIVE
ACCESS
TECHNOLOGIES

Gigabit Ethernet Access Architectures

Even though estimates of the rate of increase vary, the increasing importance of IP-based traffic is certainly undeniable. Today, the primary network type for transporting IP packets effectively within an enterprise is Ethernet. Therefore, it is certainly no surprise that Ethernet is envisioned to find a place in public networks. Because the cost and manageability of Ethernet networks has been improving for years, a number of service providers are starting to feel comfortable about this technology as a service provider platform and take advantage of Ethernet in metropolitan-area and wide-area networks.

The availability of dark fiber networks has allowed several new service providers to begin offering native Ethernet-based services to businesses at attractive prices. In most cases, the network chosen for Gigabit Ethernet services is a simple point-to-point fiber or a wavelength on a fiber, with either a switched mesh or a metropolitan-area ring architecture to connect multiple sites to a long-haul carrier's point of presence. However, the deployment of native Ethernet services is still in the early stages. Nevertheless, it is expected that this type of service will become a significant factor in networks in coming years.

Because traditional Ethernet has not been used in public networks until recently, it is no surprise that, unlike in traditional architectures such as SONET, the requirements of most service providers for redundancy, reliability, latency, and equipment packaging have not been addressed by previous generations of equipment.

A number of technical innovations are on the horizon to address some of these difficulties. One of them is led by the Resilient Packet Ring (RPR) Consortium, which includes a number of vendors working to define standards for ring-based Ethernet, with the goal of providing Ethernet with the reliability but not the complexity of SONET (see Figure 9.19).

Vendors and system architects are also investigating the possible combination of Ethernet-based PON systems for access in combination with RPR, Gigabit Ethernet, or 10-Gigabit Ethernet systems as transport platform. Some forward-looking housing and business park developers started to lay fiber into the ground and teamed up with service providers and equipment vendors when they were building new business parks and residential communities. These environments provide a good experimental platform to try new services. Information regarding these efforts can be found at http://www.ruralfiber.net/ or http://www.pa-fiber.net/.

Table 9.4 summarizes the various fiber access technologies that have been discussed in this section.

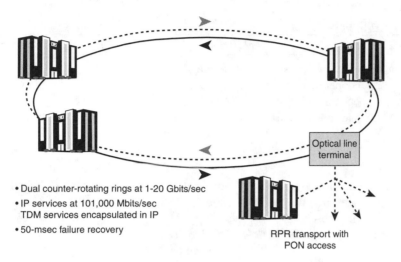

- Dual counter-rotating rings at 1-20 Gbits/sec
- IP services at 101,000 Mbits/sec
  TDM services encapsulated in IP
- 50-msec failure recovery

Optical line
terminal

RPR transport with
PON access

FIGURE 9.19

Ethernet Resilient Packet Ring (RPR) architecture.

TABLE 9.4 Comparison of Optical Fiber Access Technologies

Technology	Description	History	Weaknesses	Strengths
Fiber to the curb	Extension of fiber close to the customer, typically within less than 1,000 feet with copper or coaxial cable to remainder	Derived from widely deployed digital-loop-carrier architecture—maybe TDM or ATM	Many remote terminals not capable of easy upgrade	Excellent for voice; data in low density–video requires overlay
Hybrid fiber coaxial systems	Extension of fiber close to the customer, typically less than 3,000 feet, then coaxial cable to remainder	Evolution of widely deployed cable TV fiber-node architecture, typically frequency-division multiple access, with digital/analog	Limited upstream bandwidth in radio frequency; noise ingress; conventional TDM voice services difficult	Excellent for video; very low cost deployment to subscriber

9

TABLE 9.4 Continued

Technology	Description	History	Weaknesses	Strengths
Passive optical network	Fiber or fiber-copper star, using optical splitting and time-division multiple access	Emerging architecture— maybe ATM, IP, or proprietary transport	Splitter technology still unproven in outside plant; capacity planning not easy	Eliminates outside plant electronics; saves feeder fiber; adapts well to subscriber growth scenarios
Direct point-to-point or collapsed ring	Star topology, dedicated fiber per subscriber	Traditional architecture, typically using SONET or TDM; now being applied to IP services on fiber (typically Gigabit Ethernet)	Low fiber utilization; for Gigabit Ethernet, redundancy and reliability are limited; voice solutions still emerging	Excellent for native rate LAN extension services; bandwidth provisioning very easy

Besides all the excitement regarding fiber in the local loop and broadband access, it is important to remember that the fiber deployment in these environments is basically in its infancy. This is despite the fact that the benefits of fiber in the local loop have been promoted and extensively discussed for more than a decade. Up to now, a lower risk business model has not existed that could justify the tremendous cost associated with laying a new infrastructure into the ground. As a result, only about 5% of the commercial buildings in the major U.S. metropolitan areas rely on fiber. No reliable figures are available regarding the deployment of fiber in business parks and residential areas outside the metro. However, even the most optimistic fiber proponents agree that this figure is much lower than the figure given for metropolitan business buildings.

FSO Versus the Alternatives

The previous sections discussed various copper, wireless, and fiber-based access strategies commonly used to bridge the local loop access bottleneck. No technology fulfills all requirements imposed by the service provider and the end user. Without a doubt, technologies based on optical fiber provide the necessary bandwidth. However, the implementation of local loop fiber networks is time consuming and extremely costly. Alternative approaches using either the existing wire-based infrastructure or wireless access solutions are less expensive and easier to implement. However, these strategies do not provide the bandwidth to satisfy growing customer demands for higher speed services.

The rapid adoption of the Internet over the past decade creates an "optical dead zone" between high-capacity long-haul networks and local loop access networks that are dictated by the need for higher bandwidth. The mix of broadband access technologies commonly used in today's networks is summarized in Table 9.5. Optical communication technology is certainly the most promising approach as far as high-speed access is concerned. FSO is certainly capable of providing fiber speed access, but without the long time delays caused by laying fiber and also without the enormous cost of deploying fiber networks.

TABLE 9.5 Commonly Used Broadband Access Technologies

Technology	Accelerators	Inhibitors
Ethernet-based point-to-multipoint	Simplicity, affordability, Gigabit-plus bandwidth	Evolving standards
ATM-based point-to-multipoint	Mature standards	Bandwidth limitations, costly equipment
Ethernet-based point-to-point	Simplicity, Gigabit-plus bandwidth	Customer acquisition/capital expenditure mismatch
T1/E1	Proven track record of service	Problematic provisioning, limited bandwidth
DSL	Affordability	Distance limitations, problematic provisioning
Broadband cable	Affordability, high bandwidth	Network retrofit, shared bandwidth, lack of business penetration
Fixed wireless	High bandwidth	Line-of-sight requirements
Free-space optics	High bandwidth, less problematic provisioning	Distance limitations between nodes

FSO systems have limitations in distance, and like any other ultra-high frequency technology, FSO systems require line of sight between networking locations. However, distances in local loop access environments are typically less than a mile; therefore, the limitation in distance might not be a real constraint.

Another feature that makes FSO attractive when compared to high-bandwidth fixed wireless microwave solutions is the fact that FSO systems do not need a license for operation. In addition, in contrast to unlicensed wireless systems, FSO systems are extremely secure and are not subject to electromagnetic interference. Because FSO systems can be designed to operate as physical Layer 1 systems, FSO simply provides a connectivity pipe rather than a protocol- and topology-dependent access strategy. This is the reason that FSO systems are sometimes referred to as fiber optics systems. Service providers starting to deploy FSO systems envision FSO as a fiber augmentation or a fiber replacement strategy rather than an access strategy. Similar to a piece of optical fiber, FSO represents physical layer connectivity; therefore, it is perfectly suited to accelerate the deployment of access networks, while integrating into existing infrastructure.

Summary

At the present state of technology development, FSO is well suited to fill gaps in metropolitan loop access networks. FSO will be used to close gaps in metropolitan area networks, such as ring closures or mesh completions.

Undoubtedly, FSO will migrate further into access networks that are located outside of the traditional business-oriented metropolitan fiber networks.

The Outlook for FSO

IN THIS CHAPTER

One thing is clear, and that is the inevitable growth in our networks in terms of both users and applications. The number of users is on the rise, and applications are becoming more bandwidth intensive. The Internet is partly responsible, but so is the change in the communication culture. Either way, existing networks are not keeping up with the change. This chapter looks briefly at how these challenges signal good things for the future of FSO.

Service Providers, Business Customers, and Residential Customers

For service providers, speed in addressing bandwidth challenges is critical to their business success. A failure to address these needs quickly could result in loss of revenues, customers, and market share. So, fast is the mantra by which the service providers should operate. FSO's extremely fast deployment characteristic should be found to be quite beneficial by service providers in addressing these needs.

These service providers must remember that while they are acting quickly, they must also consider cost. Merely addressing the needs will not address the needs of their customers—optical capacity at a decreasing cost per bit.

An ideal solution to these problems and issues is the deployment of free-space optics. Service providers will benefit tremendously by deploying this technology as enterprises historically have. FSO is now ripe for deployment in the networks of service providers worldwide. The outlook for FSO is positive, and it is even more valuable for service providers who are innovators and fast movers.

The integration of FSO in optical networks was discussed extensively in Chapter 4, "Integration of FSO in Optical Networks." The bandwidth requirement in these kinds of environments is predicted to increase continually within the next decade. This demand is dominantly driven by video and data applications. Free-space optics has the potential to be part of the technology platform to satisfy this ever-growing bandwidth demand.

Today, and undoubtedly in the future, businesses will be the main consumers of high bandwidth. This will drive the need for "optical" bandwidth closer to the edge of metropolitan and campus networks. The vision of forward-looking network designers clearly points in the direction of an all-optical network implementation in these kinds of environments. Due to the similarity in the transport mechanism between FSO and fiber optics, these two technologies can be used in a synergistic way to accomplish this task.

From the consumer side, data-intensive end user network applications, such as downloading of video/audio titles, online remote teaching, or interactive gaming, will drive the need for higher bandwidth to the edge of residential areas. As discussed in Chapter 9, "Alternative Access

Technologies," the technology restrictions of a copper-based infrastructure will hardly be able to keep up with the demand in the nearest future. FSO might become the emerging technology to open residential communication bottlenecks.

Moving to the Edge and Residential Areas

People live either in multitenant/multidwelling units (MTU/MDU) or single family homes. MTUs are especially popular in densely populated environments that offer limited space, such as Japan. In contrast, a country such as the U.S. has a fair number of people living in single homes clustered in multiple-home communities or smaller and fewer unit townhouse constructions. The following sections describe how FSO can be used or play an important role within the communication infrastructure in residential areas.

MTU/MDU Networks

There are about 28 million multitenant units in North America. Of these 28 million buildings, 8 million have more than 10 units that could benefit from and pay for the delivery of higher-speed data and video-based broadband services. With respect to potential buildings to be served, this market is certainly much bigger than the market for business buildings, which count for 800,000 buildings in North America.

However, the business model around the MTU residential market is quite different from the business customer end user market. Whereas businesses typically have the financial strength to pay for higher-speed services, the residential MTU market lives under a much more constrained financial service model. In today's environment, this market is dominated by providing service through the copper-based telephone or cable TV infrastructure.

In the residential market, and opposite to the service market, cable modem services have a much higher penetration rate than DSL services. This largely reflects the fact that DSL services in residential areas are constrained by the distance limitation between the MTU and the nearest central office (CO). Cable modem services through an HFC infrastructure are provided from the nearest cable head end unit that is typically closer to the end user in residential environments. However, during the past couple of years, DSL providers constructed "mini COs" that are located closer to the end user.

Both cable modem and DSL service providers especially take advantage of the existing in-building copper infrastructure to reach the end user either through the cable TV or the telephone outlet. Unlike business buildings, the construction of a new indoor distribution plant—such as fiber-based optics in building risers—is simply not feasible from a cost prospective.

The balance between costs of provisioning services and the potential service income stream is one of the main concerns in the business model around MTU residential access. However, with more and more people signing up for high-speed services in the MTU market, bandwidth bottle-necks due to oversubscription are already obvious now. If a building with 10 end users who are guaranteed a bandwidth of at least 1 Mbps is served and shared by a 1.544 Mbps T1 line back-bone connection, a drastic speed degradation will be unavoidable during peak traffic hours. Increasing the backbone bandwidth capacity to a multitenant building is the only solution to keep the end user happy and guarantee the long-term success of this business model. In this scenario, FSO can play an important role. The distances between MTUs are typically short, and this certainly benefits the FSO deployment strategy. In addition, the cost of deploying FSO versus fiber will certainly benefit the cost-sensitive MTU business model.

From the network design point of view, this deployment suggests running fiber to more of the MTU buildings or tapping into the network of the closet CO, if possible. From there, place an appropriately sized DSLAM into each of the nearby MTU buildings. Most of the newer DSLAM equipment provides either a higher-capacity 100 Mbps Ethernet or a DS-3/OC-3 interface. These interfaces can be connected directly to an FSO link located either on the roof or behind a window of the MTU building. Multiple buildings can be interconnected in a combination of a mesh, tree, star, or hub and spoke topology. The higher capacity of the FSO system when com-pared to a copper-based T1 line will greatly eliminate the problem of massive user oversub-scription. It will also allow the service provider to offer new services that require higher bandwidth. Figure 10.1 depicts an FSO design for an MTU network.

For Ethernet-based services or in environments where the internal copper-based infrastructure is either inaccessible or not available to the service provider, the FSO backbone approach could be coupled with another unlicensed wireless access strategy such as 802.11a or b, HyperLAN, or UNII band. The security aspect of using a wireless in-building distribution system can be addressed by limiting the transmission power and encryption of the data stream. This approach would allow independent service providers to build a complete "bypass" network and, in addi-tion, provide the end user with the benefit of a wireless indoor LAN.

SDU Networks

From the service provider's point of view, the high-speed residential single-family home (sin-gle dwelling unit, or SDU) or townhome environment is much more challenging than the MTU residential market. On one hand, the total amount of revenue that must be collected from a single resident or family must be relatively comparable to lower performance solutions such as DSL. On the other hand, the deployment cost for connecting a single dwelling unit to the network is potentially not very different from the deployment cost of a multitenant unit, where the costs can be spread among many users. This is the most important aspect that limits the deployment of a high-speed infrastructure today and most likely in the near future.

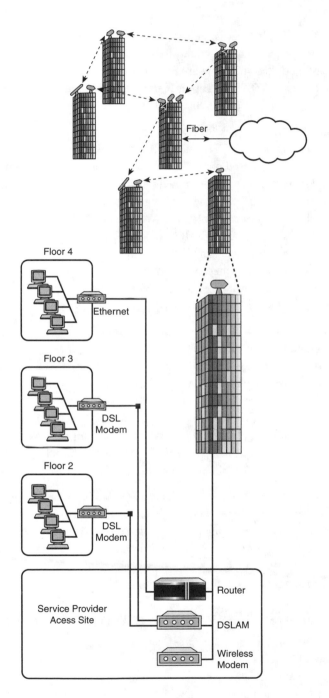

Figure 10.1

An FSO multitenant unit (MTU) network.

The key to success to the high-speed broadband residential network is closely connected to the aspect of deployment cost. It is unlikely that fiber deployment to the home will migrate faster than it did in the past, and these numbers are certainly not encouraging. The high cost of deploying fiber to the home (FTTH) is certainly the main reason that enthusiastic expectations of FTTH projections made in the early '90s were dampened.

Considering the length of time it took for phone companies to connect every home to a phone line is helpful in predicting when FTTH will become a reality. In some geographical regions where the deployment of wire-based phone lines was more or less abandoned when the cell phone started its phenomenal success, wire-based phone line infrastructure might never become reality. However, in industrialized countries, the wire-based phone and cable TV infrastructure is the foundation of the network access platform to residential buildings. The closest fiber ever came to residential buildings was through HFC and DLC networks (see Chapter 9). This will change on a large scale in the future if an alternative high-bandwidth access technology is not developed that relies on deploying fiber to each home. To be successful in this market, the deployment of this technology must be extremely inexpensive to satisfy the cost constraints in the service provider business model.

The only technology approach that could potentially enable these kinds of high-speed residential networks is a combination of the existing FTTN (fiber to the neighborhood) or FTTC (fiber to the curb) infrastructure and FSO for the "Last 100 meter" access to the home. This is the idea of the hybrid-fiber-laser (HFL) residential network. Small and inexpensive towers could be erected strategically in a typical neighborhood and serve a whole cluster of houses as an FSO distribution point. The individual homeowner could point a small footprint transceiver device toward the tower and get access to the network. The multiple towers in a larger community could be interconnected by high-capacity FSO systems. Light poles, owned by the community, could play an important role in this scenario.

To make the idea of the HF residential network a commercial success, the price for FSO transceiver equipment delivery—such as access speeds up to 100 Mbps—has to come down drastically. On the positive side, FSO equipment uses standard optical components that are used in the mass market for optical communications, and prices have decreased drastically over the past couple of years. Therefore, there is no reason to believe that FSO equipment will not be able to meet these price points in the foreseeable future.

Residential MANs: All-Optical Networks with FSO

In the near future, the hybrid, fiber-laser infrastructure could be used to build residential MANs. The residential MAN is the exact image of today's metropolitan area network on a different bandwidth scale. Whereas the capacity requirements for today's business buildings can easily scale beyond 1 Gbps and beyond and with potential to reach 10 Gbps in the near future, a residential MAN that connects MTUs or SDUs typically does not require this amount of bandwidth capacity at a single location. However, if video services, such as uncompressed high-definition television (HDTV), become part of the services offering, the sustainable bandwidth requirement might go well beyond 50 Mbps.

The benefit of this network architecture is its relative simplicity. If all services are carried in the optical layer, a network becomes highly reliable because the number of electronic network elements is reduced greatly, eliminating numerous points of failure. In addition, users have dedicated, guaranteed optical bandwidth through the network; therefore, concerns associated with oversubscription of typical data services are allayed. This network also has the benefits of bit rate and protocol transparency, and can easily migrate from one service type to another because of the agnostic quality of the optical network.

For all optical networks to be reality, the network must be all optical, end-to-end. End-to-end means that traffic from an end user at one end travels all optically to the user at the other end without an O-E-O conversion. This is not an easy task because the core, access, and edge are not completely optical. The good news is that they are moving in that direction, and free-space optics is at the cornerstone of this optical renaissance. Free-space optics will accelerate the all-optical networks and enable service providers to deploy these networks faster and more cost effectively.

Environment and Community

You have seen some effects of technology that have impacted the balance of ecosystems. In a quest for a technology to make life easier, it is equally important not to impact the natural balance of ecosystems. Similarly, communities and technology should be in friendly association. Preservation of communities and addressing their basic needs without destroying their core is a key factor in successful technology deployment.

Free-space optics address both the need for communal integrity and environmental balance.

FSO offers badly needed pain relief for cities where torn-up streets and disruption are commonplace. As telecommunications carriers race to lay high-speed fiber-optic cable in the ground, cities are facing new and unforeseen challenges. Recently, U.S. cities wrestled with difficult issues such as street closures, broken water mains, and merchant upheaval. Some cities have instituted moratoriums or restrictions on the installation of fiber-optic cable to reduce disruption to their business districts.

Digging up streets causes more than physical damage to communities. Torn-up streets result in increased traffic, which means that more pollution contributes directly to ozone depletion. Digging up streets and medians leads to the displacement of trees. In older communities, such disruption could destroy pieces of history that might never be recovered again. Who wants a disruptive technology at this price? The costs and associated pain of fiber trenching are not ideal for business or the environment. With FSO, you can have high bandwidth at rates that are cost-effective and eco-friendly.

The Competitive Landscape

Two years ago only a handful of companies were offering FSO systems. A lot has changed since then. With continued acceptance of FSO as a viable, cost-effective alternative to fiber-optic cable, an increasing number of players have entered this space.

Most of these companies offer low-end systems to address the enterprise market needs. A select few offer products that have the potential to address carrier requirements, including LightPointe, Fsona, and Optical Access. Companies use different approaches. For example, LightPointe's approach is to be a Layer 1 vendor of FSO equipment, mainly enabling service providers to deploy point-to-point systems. Airfiber, on the other hand, markets a local loop architecture, an ATM-based mesh network using FSO as the enabling technology. Terabeam is yet another example of how FSO is being used. Terabeam's approach is to be a service provider and offer broadband services using FSO in a point-to-multipoint topology. All three approaches are valid, but the winners in the space will be the ones who address the needs of service providers to enable optical services and generate revenue quickly whether through a point-to-point, mesh, or point-to-multipoint approach.

Summary

The networks are moving to optics. Light is fast becoming the medium of connectivity, driven by its versatility and flexibility in carrying diverse information. It is further strengthened by the low costs associated with it. It is a medium of connectivity, but a medium of transport is also needed. Fiber-optic cable is the ideal medium of transport—a glass tunnel through which information travels at the speed of light, connecting this world optically.

Unfortunately, this medium of transport does not reach most end users because of the high costs associated with laying fiber and the lack of guarantees for continued business. So how is optical connectivity enabled?

The answer is free-space optics. Free-space optics will enable service providers to extend their optical reach. FSO is an emerging technology with the dramatic benefits of low cost, flexible and quick deployment, the promise of optical bandwidth, and the economies of scale strongly desired by service providers.

The following can be predicted about FSO's future:

- Large-scale adoption by carriers in the next couple of years
- Decreasing cost per bit per mile
- All-optical connectivity using free-space optics
- FSO becoming part of the network toolkit of service providers

In addition, you will likely see improved distances and availabilities and the extension of high bandwidth to the home using FSO. The future of FSO looks promising, but its success largely depends on its large-scale adoption by the service providers and resulting shift from a niche application to a mainstream technology.

Will FSO move from niche to mainstream? The transition has already begun.

Frequently Asked Questions

Q: What is Free-Space Optics?

A: Free-Space Optics (FSO) is an optical technology that uses beams of light over air (instead of fiber-optic cable) to deliver reliable, high-speed optical bandwidth connections both cost-effectively and quickly. FSO products can be deployed in hours versus months, without the expense and hassles of digging up roads for cable and applying for FCC licenses.

Q: Who uses FSO?

A: An estimated 93% of all businesses are within a mile of fiber-optic cable but don't touch fiber. Bandwidth demands continue to explode, but lead times for installing fiber-optic cable average 14 months.

FSO allows carriers to grow fiber-optic networks without the cost constraints and the bureaucratic red tape necessary to obtain trenching permits. For cash-strapped service providers, FSO technology provides a proven route to quickly gain new customers and revenue.

Q: Are all FSO products transparent?

A: No, not all products are transparent. Some are ATM based whereas others are IP based. Some products, such as LightPointe and Fsona, are transparent because they take a Layer 1 approach to FSO.

Q: In which type of topologies can FSO be deployed?

A: FSO can be deployed in multiple topologies ranging from point-to-point, mesh, star, rings, or point-to-multipoint topologies. Point-to-point forms the underlying technology for most of the topologies.

Q: Is FSO only deployed on rooftops?

A: No, FSO can be deployed behind windows, on rooftops, or on a combination of both. It depends on the vendor and its approach.

Q: What services can be carried through FSO links?

A: A wide variety of telecom/datacom protocols can be carried over these systems. Some equipment is protocol agnostic and carries transparently any signal sent through a fiber. Some use SONET, ATM, or Ethernet protocols. The usual E1/T1 lines, T3, OC3, and OC12 data rates can be delivered.

Q: What are the typical FSO applications?

A: FSO products can be used to provide optical connectivity in multiple applications, such as metropolitan network extension, DWDM services, access/last mile, wireless backhaul, disaster recovery, storage area networks, and LAN solutions, among others, addressing needs of both carriers and enterprise customers.

Q: What speeds can FSO products offer?

A: Ranges of bandwidth starting at 1 Mbps–2.5 Gbps are available today. Shortly, products of 10 Gbps will be available.

Q: Is FSO safe?

A: Yes, FSO products are eye safe and environmentally safe. Most products meet or exceed standards set by U.S. and international regulatory bodies. For more information on safety, please refer to Chapter 7, "Free-Space Optics and Laser Safety."

Q: What are the price-point advantages to using Free-Space Optics?

A: Connecting to fiber typically requires access fees ranging from $200 to $20,000 per month, depending on the size and scope of the network. Trenching for fiber requires permits, time, and construction costs that can range from several thousands of dollars to a few hundred thousand dollars per mile. In comparison, FSO involves a one-time installation between $5,000 and $45,000 (no monthly recurring costs).

Q: What's the installation process? Does it take long?

A: The installation process is simple. It includes obtaining a site survey, installing the equipment, aligning the link heads, and connecting the master link to a physical fiber-optic backbone (line). The entire process takes as little as two–four hours. Please refer to Chapter 6, "Installation of Free-Space Optical Systems," for details.

Q: What are the costs of implementation and maintenance?

A: FSO products range from $5,000 to $70,000 per system (two links), depending on bandwidth and distance. The products typically do not require maintenance, other than occasional realignment.

Q: Are special tools required for installation?

A: Only a few simple hand tools and drills are required to install most FSO units. The links are equipped with a binocular or a camera for coarse alignment. For final adjustment, most systems are equipped with an acoustic tone or an optical power level meter. Automatic tracking systems are also available.

Q: Is FSO technology a temporary solution?

A: Yes, FSO can be used as a temporary solution. However, IP traffic will continue to drive demand for bandwidth. Early customers—enterprise end users—do not plan to replace their links with fiber connections. One customer, for example, says he will save an estimated $500,000 over five years by using FSO versus traditional fiber-optic cable.

Q: Do you need licenses to operate FSO products?

A: No licenses are required for FSO like they are for wireless. FSO operates in the unregulated spectrum.

Q: What are the operating wavelengths?

A: Most current systems are infrared; depending on the customer's requirements for speed and range, FSO uses either high-power light-emitting diodes (LEDs) or high-power laser diodes (LDs). The operating wavelength is in the near-infrared region of the electromagnetic spectrum at a wavelength around 850 nm and 1550 nm.

Q: What are the advantages of using infrared communication instead of other wireless technologies?

A: Infrared communication offers a much higher bandwidth than other wireless solutions, such as spread spectrum or microwave links. Infrared communication does not require FCC approval. Infrared technology is jamming resistant and has a much higher signal security than other wireless solutions.

Q: What is the recommended range of the FSO product?

A: This depends on the speed and weather conditions. Typical links are between 300 m and 4 km, although longer distances such as 9–11 km are possible depending on the speed and required availability.

Q: What are the power requirements?

A: FSO terminals typically require a voltage of 115 or 230Vac.

Q: How are the links connected to the network?

A: The connection to the network is accomplished by using two optical fibers (send and receive)—multimode or single mode—with standard (usually ST or SC) fiber connectors. Some vendors also provide electrical interfaces such as E1/T1.

Q: What is the physical size and weight of a unit?

A: The product sizes are variable, but they typically weigh between 10–25 lbs.

Q: How secure are FSO transmissions?

A: Optical transmission of data through air is one of the safest transmission methods. Due to the narrow beam of the systems (approximately 2–6 mrad), it is virtually impossible to tap into the free-space optical connection without interrupting the beam path. Because the wavelength of the signal is in the infrared range that is invisible to the human eye, it is also difficult to fix a position for the beam. Anyone or anything tapping into the communication path would have to be mounted either between or behind the actual free-space optical link heads. The former is unlikely because the mounting height of the system is always greater than 5–10 m (to prevent signal interrupts from automobiles or persons); therefore, a stable pole would be required whose presence in the beam path could not be overlooked. Free-space optical technology has been around for many decades and has been successfully used in high security applications (mainly defense and space).

Q: How does weather affect performance?

A: The link margin of a typical system is about 30 decibels (dB). This value is important because rain, snow, fog, and so on are changing the attenuation of the atmosphere. A rainfall of 1 inch/hour roughly corresponds to an attenuation of 7 dB/km. At a 1 km range, the systems will operate under the following conditions: a rain rate of about 4 inches per hour, a wet snow rate of less than 2 inches per hour, and a dry snow rate of less than 1 inch per hour. In fog, and as a simple rule of thumb, the visibility should be greater than 90% of the link distance to ensure uninterrupted availability of the system.

Q: What effect does sunlight have on the link?

A: FSO systems use narrowband optical filters to minimize the effect of direct sunlight. If possible, the laser systems should not be mounted under a steep angle in a direct East-West orientation to avoid the effects of direct sunlight. Direct sunlight into the front of the unit can result in short periods of time when the receiver will be inoperable due to saturation of the receiver photo diode. These outages can last for several minutes depending on the time of the year and the angle of the sun in the sky. However, due to the narrow reception angle of the receiving optics, the sun must appear almost directly behind the link head. Therefore, sunlight is potentially a problem only if one of the link heads points under a steep angle into the sky. The system will fully recover after the sun is out of the angle of view of the receiver. Most vendors incorporate narrowband sunlight blocking filters that drastically minimize the impact of direct sunlight.

Q: What effect does scintillation (heat shimmer) have?

A: Scintillation or heat shimmer imposes a low-frequency variation on the amount of light detected by the receiver. If the amount of light detected falls below the required light threshold of the receiver, short bursts of errors will occur. Networks such as Ethernet and Token Ring will retransmit this lost data. The systems are designed to minimize the effects of scintillation. Proper site and optical path selection can eliminate the effects of scintillation entirely as well as the use of a multibeam system.

Q: Can FSO systems operate through glass?

A: Infrared links can operate through glass; however, for each glass surface, the light intensity will be reduced by approximately 4%. The glass should not be coated with an infrared reflecting or absorbing material because all light might be lost. As the angle of the beam with the glass increases, more light is reflected until the critical angle of approximately 42 degrees is reached. Above the critical angle, all of the light is reflected off the glass and no signal can reach the receiver.

Q: What happens if a bird flies through the beam?

A: In most cases, the beam is too wide to be interrupted by a bird. In exceptional cases, if a bird were to fly close to the link transceiver, a momentary interruption might occur, causing a short burst of errors. In Ethernet or Token Ring networks, the corrupted data packages will be retransmitted and the user will most likely not see an effect. Additionally, some manufacturers have multibeam equipment, which transmits and receives information by using multiple beams. The probability that a bird will interrupt all beams at the same time is extremely low.

Laser Safety Resources

IN THIS CHAPTER

This appendix lists references for more complete information on laser safety standards.

Safety and Compliance Standards for Manufacturers

A comprehensive resource on laser safety standards issued by the U.S. Military, U.S. Federal Government, various commercial agencies, and international-multinational agencies is available at `http://www.navylasersafety.com/standards/standards.htm`.

Following is a run-through of the most pertinent standards, and additional Web sites for retrieval of the documents.

21 CFR 1040, *Laser Product Performance Standard,* U.S. Center for Device and Radiological Health (CDRH). Regulations are mandatory for all laser products sold to end users in the United States. This document is available from the CDRH (`http://www.fda.gov/cdrh/radhlth/`).

IEC 825-1, *Safety of Laser Products—Part 1: Equipment Classification, Requirements, and User's Guide.* International Electrotechnical Commission (IEC). Published in 1993 and amended in 1997, it is being redesignated as IEC 60825-1.

For convenience, it is divided into three separate sections:

- Section One (general) and the annexes
- Section Two (manufacturing requirements)
- Section Three (User's Guide)

This provides requirements for manufacturers that are similar to the CDRH laser regulations and user requirements that are similar to those in the ANSI Z136.1 laser safety standard. In English and French; 100 pages. This document is available from ANSI (`http://www.ansi.org/`) in New York and from the IEC (`http://www.iec.ch`) in Geneva.

EN60825-1, *Safety of Laser Products—Part 1: Equipment Classification, Requirements, and User's Guide.* Cenelec. Published in 1994 and amended in 1996, and essentially identical to IEC 825-1. Available in English from the British Standards Institute (BSI) (`http://www.bsi-global.com/`) in London.

IEC 60825-1, Amendment 2 (2001), *Safety of Laser Products—Part 1: Equipment Classification, Requirements, and User's Guide*. This is the latest amendment of the IEC laser safety standard and the most recent and relevant standard regarding laser eye safety. The IEC adopted this new classification standard as of March 1, 2001 in all countries that are covered under the IEC regulation. CDRH has committed to unifying its compliance standards with those established by the IEC in the near future.

Laser Safety Standards for Users

ANSI Z136.1, *Standard for the Safe Use of Lasers*. The basic laser safety document for users of laser products, including manufacturing facilities. Available from the Laser Institute of America (LIA) (`http://laserinstitute.org/`).

IEC 825-1, *Safety of Laser Products—Part 1: Equipment Classification, Requirements, and User's Guide*. International Electrotechnical Commission (IEC). Published in 1993 and amended in 1997. Is being redesignated as IEC 60825-1. For convenience, it is divided into three separate sections: Section One (general) and the annexes; Section Two (manufacturing requirements); and Section Three (User's Guide). This provides user requirements that are similar to those in the ANSI Z136.1 laser safety standard. In English and French; 100 pages. Available from ANSI (`http://www.ansi.org/`) in New York and from the IEC (`http://www.iec.ch`) in Geneva.

IEC 60825-1, Amendment 2 (2001), *Safety of Laser Products—Part 1: Equipment Classification, Requirements, and User's Guide*. This is the latest amendment of the IEC laser safety standard and the most recent and relevant standard regarding laser eye safety. The IEC adopted this new classification standard as of March 1, 2001 in all countries that are covered under the IEC regulation. CDRH has committed to unifying its compliance standards with those established by the IEC in the near future.

Laser Safety Standards Organizations

CDRH (Center for Devices and Radiological Health)—An agency within the U.S. FDA that publishes and enforces legal requirements on lasers.

IEC (International Electrotechnical Commission)—An organization that publishes international standards on electrical subjects. These are not laws, and the adoption and enforcement of IEC standards are at the discretion of individual nations.

B

LASER SAFETY
RESOURCES

ISO (International Standards Organization)—An organization that is equivalent to the IEC, except that the ISO publishes international standards on nonelectrical subjects.

CEN and CENELEC—European equivalents of ISO and IEC. CEN and CENELEC standards are typically European Norms (EN), and many are published in response to directives from the European Commission.

ANSI (American National Standards Institute)—A U.S. organization that publishes standards for laser users. The ANSI Z136.1 general laser safety standard is not a law, but it forms the basis for state and OSHA requirements for the use of lasers. Other standards in the ANSI Z136 series are intended for specific applications. An ANSI B11 committee publishes standards for machine tool safety.

LIA (Laser Institute of America)—An organization that provides laser safety information, including conferences, symposia, publications, and training courses. Publications include the ANSI Z136 series of laser safety standards and the *Journal of Laser Applications*.

Glossary

acceptance angle   The half angle of the cone within which incident light is internally reflected by the fiber core. It is equal to Arcsin (NA). In FSO systems this value is often used to define the receive optics field of view.

amplitude modulation (AM)   A means of signal transmission whereby transmitter (light source) signal intensity is varied in relation to the amplitude of the input signal.

analog   A format that uses continuous physical variables such as voltage amplitude or frequency variations to transmit information.

angle of incidence   The angle between an incident ray and the perpendicular angle to a reflecting surface.

APD (Avalanche Photodiode)   A photodiode designed to take advantage of avalanche multiplication of photocurrent. As the reverse bias voltage approaches the breakdown voltage, hole electron pairs created by absorbed photons acquire sufficient energy to create additional hole electron pairs when they collide with ions; thus, a multiplication or signal gain is achieved.

aramid yarn   Strength element used in Siecor cable to provide support and additional protection of the fiber bundles. Kevlar is a particular brand of aramid yarn.

armoring   Additional protection between jacket layers that provides protection against severe outdoor environments. Usually made of plastic-coated steel, and can be corrugated for flexibility.

attenuation   (1) Limited operation. The condition in a fiber-optic link when operation is limited by the power of the received signal rather than by bandwidth or by distortion. (2) The decrease in magnitude of power of a signal in transmission between points. A term used for expressing the total losses on an optical fiber consisting of the ratio of light output to light input. Attenuation is usually measured in decibels per kilometer (db/km) at a specific wavelength. The lower the number, the better the fiber. Typical multimode wavelengths are 850 and 1,300 nanometers (nm); single-mode wavelengths are typically 1,300 and 1,500 nm. When specifying attenuation, it is important to note if it is a nominal or average room temperature value or a maximum overoperating range. (3) In FSO systems this term is used to describe the impact of the atmosphere.

attenuator   A passive optical component that intentionally reduces the optical power propagating in a fiber.

average power   The average level of power in a signal that varies with time.

axial ray   A light ray that travels along the axis of an optical fiber.

back reflection   Connector Physical Contact (PC) connectors provide a back reflection characteristic exceeding <-30 dB. Super Physical Contact (SPC) connectors provide a back reflection characteristic exceeding <-40 dB. Ultra Physical Contact (UPC) connectors provide a back reflection characteristic exceeding <-50 dB. Angled Physical Contact (APC) connectors provide a back reflection characteristic exceeding <-60 dB. *See also* reflectance.

backscattering   A small fraction of light that is deflected out of the original direction of propagation by scattering and suffers a reversal of direction. In other words, this is light propagated in the optical waveguide or in FSO systems toward the transmitter.

bandpass   A range of wavelengths over which a component will meet specifications.

bandwidth   The information-carrying capacity of of the transport media. Expressed in MHz-km, the bandwidth value specifies the analog bandwidth capability or number of digital transitions per second that the transport media can sustain over a 1 km distance. Bandwidth is dependent on wavelength and type of light source.

bandwidth-limited operation   The condition prevailing when the system bandwidth, rather than the amplitude of the signal, limits performance. The condition is reached when modal dispersion distorts the shape of the waveform beyond its specified limits.

baseband   A method of communication in which a signal is transmitted at its original frequency without being impressed on a carrier.

BAUD   A unit of signaling speed equal to the number of signal intervals per second, which might or might not be equal to the data rate in bits per second. In some encoding schemes, such as Non Return to Zero (NRZ), BAUD equals the data rate. In others, such as Manchester encoding, two transitions per bit are required.

beamsplitter   A device used to divide an optical beam into two or more separate beams.

bend loss   A form of increased attenuation in a fiber that results from a fiber bending around a restrictive curvature (a macrobend) or from minute distortions.

bend radius   (1) The radius that a fiber can bend before it breaks or increases in attenuation. (2) Cable bend radius.

BER   Bit error rate. Specifies expected frequency of errors. The ratio of incorrectly transmitted bits to correctly transmitted bits.

bit   A binary digit, which is the smallest element of information in binary system. A 1 or 0 of binary data.

bps (bits per second)   The number of energy pulses passing a given point in a transmission medium in one second.

break-out cable   Multifiber cable constructed in the tight buffered design. Designed for ease of connection and rugged applications for intra- or interbuilding requirements.

broadband   The ability of a system to carry a multitude of signals simultaneously. In data transmission, it denotes transmission facilities capable of handling frequencies greater than those required for high-grade voice communications. The higher frequency allows the carrying of several simultaneous channels. Broadband infers the use of a carrier signal rather than direct modulation (that is, baseband).

buffer   (1) A protective material extruded directly on the fiber coating to protect it from the environment. (2) A tube extruded around the coated fiber to isolate it from

stresses on the cable. The primary buffer (next to the cladding) is 250 microns in diameter. A secondary buffer of 900 microns in diameter is used on indoor cables.

buffered fiber   Fiber protected with an additional material, such as hytrel or nylon, to provide ease in handling, connection, and increased tensile strength.

building entrance   Terminal cable entrance point where typically a trunk cable between buildings is terminated and fiber is then distributed through the building.

bundle   (1) Many individual fibers contained within a single jacket or buffer tube. (2) A group of buffered fibers distinguished in some fashion from another group in the same cable core.

bus   Commonly called *data bus*, this is a term used to describe the physical linkage between stations on a network sharing some common communication.

bus network   A network topology in which all terminals are attached to a transmission medium serving as a bus.

byte   A unit of 8 bits (digital data).

cable   An assembly of optical fibers and other material providing mechanical and environmental protection and optical insulation of the waveguides.

cable assembly   Fiber-optic cable that has connectors installed on one or both ends. General use of these cable assemblies includes the interconnection of multimode and single-mode fiber-optical cable systems and opto-electronic equipment. If connectors are attached to only one end of the cable, it is known as a *pigtail*. If connectors are attached to both ends, it is known as a *jumper*.

cable bend radius   Cable bend radius during installation infers that the cable is experiencing a tensile load. Free bend infers a lower allowable bend radius since it is at a condition of no load.

Carrier Sense Multiple Access with Collision Detection (CSMA/CD)   A technique used to control the transmission channel of a local area network to ensure that the terminals that want to transmit don't have a conflict.

CCMQJ   Certified Commercial Measurement Quality Jumper. A high-quality reference cable designed to provide accurate and consistent test results.

center wavelength(s)   The nominal operating wavelength(s).

central member   The center component of a cable. It serves as an antibuckling element to resist temperature-induced stresses. It sometimes serves as a strength element. The central member is composed of steel, fiberglass, or glass-reinforced plastic.

central office (CO)   The place where communications' common carriers terminate customer lines and locate switching equipment that interconnects those lines.

channel   A communications path or the signal sent over that channel. Through multiplexing several channels, voice channels can be transmitted over an optical channel.

chromatic dispersion   Spreading of a light pulse caused by the difference in refractive indices at different wavelengths.

cladding   The material surrounding the core of an optical fiber. The cladding must have a lower index of refraction to steer the light in the core.

cladding mode   (1) A mode confined to the cladding. (2) A light ray that propagates in the cladding.

coating   A material put on a fiber during the drawing process to protect it from the environment.

conduit   Pipe or tubing through which cables can be pulled or housed.

connector   A mechanical device used to align and join two fibers to provide a means for attaching and decoupling them to a transmitter, receiver, or another fiber. Commonly used connectors include the FC, FCPC, Biconic, ST Connector-Compatible, D4, SMA 905, or 906.

core   The central region of an optical fiber through which light is transmitted.

core eccentricity   A measure of the displacement of the center of the core relative to the cladding center.

coupler   (1) Commonly called a splitlet, it is a passive device that distributes optical power among two or more ports. It can be in various ratios. (2) A multipod device used to distribute optical power.

coupling efficiency The efficiency of optical power transfer between two components.

coupling loss The power loss suffered when coupling light from one optical device to another.

coupling ratio The percentage of light transferred to a receiving output port with respect to the total power of all output ports.

CPE CPE is an abbreviation for Customer Premises Equipment.

critical angle The smallest angle from the fiber axis at which a ray can be completely reflected at the core/cladding interface.

cutoff wavelength The shortest wavelength at which only the fundamental mode of an optical waveguide is capable of propagation.

data rate The maximum number of bits of information that can be transmitted per second, as in a data transmission link. Typically expressed as megabits per second (Mbps). dbm Decibel referenced to a milliwatt. dbp Decibel referenced to a microwatt.

dBm Output power of a signal referenced to an input signal of 1mW (Milliwatt). 0 dBm = 1 mW.

decibel (dB) Unit for measuring the relative strength of a signal. Power level referenced in decibels to a microwatt.

demultiplex The process of separating optical channels.

detector (1) A transducer that provides an electrical output signal in response to an incident optical signal. The current is dependent on the amount of light received and the type of device. (2) A semiconductor device that converts optical energy to electrical energy.

diameter-mismatch loss The loss of power at a joint that occurs when the transmitting half has a diameter greater than the diameter of the receiving half. The loss occurs when coupling light from a source to fiber, from fiber to fiber, or from fiber to detector.

dielectric Nonmetallic and, therefore, nonconductive. Glass fibers are considered dielectric. A dielectric cable contains no metallic components.

digital (1) A data format that uses two physical levels to transmit information. (2) A discrete or discontinuous signal.

directivity Also referred to as near-end crosstalk, it is the amount of power observed at a given input port with respect to an initial input power.

dispersion The cause of bandwidth limitations in a fiber. Dispersion causes a broadening of input pulses along the length of the fiber. Three major types are (a) mode dispersion caused by differential optical path lengths in a multimode fiber; (b) material dispersion caused by a differential delay of various wavelengths of light in a waveguide material; and (c) waveguide dispersion caused by light traveling in both the core and cladding materials in single-mode fibers.

distortion-limited operation *See* bandwidth-limited operation.

duplex cable A two-fiber cable suitable for duplex transmission.

duplex transmission   Transmission in both directions, either one direction at a time (half duplex) or both directions simultaneously (full duplex).

duty cycle   In a digital transmission, the ratio of high levels to low levels.

EIA/TIA   Electronics Industry Alliance/ Telecommunications Industry Association A standards association that publishes test procedures and specifications for the telecommunications industry.

electromagnetic interference (EMI)   Any electrical or electromagnetic interference that causes undesirable response, degradation, or failure in electronic equipment. Optical fibers neither emit nor receive EMI.

encoding   A scheme to represent digital ones and zeros through combining high- and low-signal voltage states.

excess loss   (1) In a fiber-optic coupler, the optical loss from that portion of light that does not emerge from the nominally operational pods of the device. (2) The ratio of the total output power of a passive component with respect to the input power.

extrinsic loss   In a fiber interconnection, that portion of loss that is not intrinsic to the fiber but is related to imperfect joining, which can be caused by the connector or splice.

fade margin   System power margin in dB that is available to counteract atmospheric attenuation due to weather impact in FSO systems.

fan-out cable   Multifiber cable constructed in the tight buffered design. Designed for ease of connection and rugged applications for intra- or interbuilding requirements.

ferrule   A small alignment tube attached to the end of the fiber and used in connectors. Generally made of stainless steel, alumina, or zirconia, used to confine and align the stripped end of a fiber.

fiber   Thin filament of glass. An optical waveguide consisting of a core and a cladding that is capable of carrying information in the form of light.

Fiber Distributed Data Interface (FDDI) A standard for a 100 Mbps fiber-optic local area network.

fiber optics   Light transmission through optical fibers for communication or signaling.

fiber-optic link   Any optical fiber transmission channel designed to connect two end terminals or to be connected in series with other channels.

FOTP   Abbreviation for Fiber-Optic Test Procedures.

FOTS   Abbreviation for Fiber-Optic Transmission System.

free space   In FSO systems the space (air) between two terminals.

frequency   The number of pulses or cycles per second; measured in units of Hertz (Hz), where 1 Hertz equals 1 pulse/cycle per second.

frequency modulation (FM)   Transmission scheme whereby information is sent by varying the frequency of an optical carrier. A method of transmission in which the carrier frequency varies in accordance with the signal.

Fresnel reflection   The reflection of a portion of the light incident on a planar surface between two homogeneous media having different refractive indices. Fresnel reflection occurs at the air/glass interfaces at entrance and exit ends of an optical fiber.

fundamental mode   The lowest order mode of a waveguide.

fusion splice   A joining of two fibers by physically fusing the two fiber ends through heat.

fusion splicing   A permanent joint accomplished by the application of localized heat sufficient to fuse or melt the ends of the optical fiber, forming a continuous single fiber.

gap loss   Loss resulting from the end separation of two axially aligned fibers.

geometric path loss   In FSO systems the transmission power loss due to divergence of the light beam.

gimbals   Mechanical platform for bam tracking.

graded index   Fiber design in which the refractive index of the core is lower toward the outside of the fiber core and higher toward the center of the core; thus the fiber bends the rays inward and allows them to travel faster in the lower index of refraction region. This type of fiber provides high-bandwidth capabilities.

ground-loop noise   Noise that results when equipment is grounded at ground points having different potential, thereby creating an unintended current path. The dielectric of optical fibers provide electrical isolation that eliminates ground loops.

hard-clad silica   A fiber with a hard plastic cladding surrounding a silica glass core.

heat shimmer   *see* scintillation.

hybrid cable   A fiber-optic cable containing two or more different types of fiber, such as 62.5 µm multimode and single-mode.

IEEE   Institute of Electrical and Electronics Engineering.

index-matching material   A material, often a liquid or cement, whose refractive index is nearly equal to the core index. The material is used to reduce Fresnel reflections from a fiber end face.

index of refraction   The ratio of light velocity in a vacuum to its velocity in a given transmitting medium. An optical characteristic (n) of a material, referencing the speed of light in that material to a vacuum.

index profile   Curve of the refractive index over the cross section of an optical waveguide.

insertion loss   The attenuation caused by the insertion of an optical component; in other words, a connector or coupler in an optical transmission system.

isolation   Also referred to as far-end crosstalk or far-end isolation. Predominantly used in reference to WDM products, it is a measure of light at an undesired wavelength at any given port.

jumper   Fiber-optic cable that has connectors installed on both ends. *See also* cable assembly.

Kevlar   *See* aramid yarn.

kilometer   1,000 meters, or 3,281 feet. The kilometer is a unit of measurement for fiber optics.

KPSI   A unit of tensile strength expressed in thousands of pounds per square inch.

laser light   Amplification by Stimulated Emission of Radiation. A device that produces coherent light with a narrow range of wavelengths.

lateral displacement loss   The loss of power that results from lateral displacement from optimum alignment between two fibers or between a fiber and an active device.

launch angle   Angle between the propagation direction of the incident light and the optical axis of an optical waveguide.

launching fiber   A fiber used in conjunction with a source to excite the modes of another fiber in a particular way. Launching fibers are most often used in test systems to improve the precision of measurements.

LED (light-emitting diode)   A semiconductor diode that emits light when forward biased to an optical signal. A device used in a transmitter to convert information from electric to optical form. It typically has a large spectral width.

light   In the laser and optical communication fields, the portion of the electromagnetic spectrum that can be handled by the basic optical techniques used for the visible spectrum extending from the near ultraviolet region of approximately 0.3 micron, through the visible region and into the midinfrared region of about 30 microns.

lightguide cable   An optical fiber, multiple fiber, or fiber bundle that includes a cable jacket and strength.

lightwaves   Electromagnetic waves in the region of optical frequencies. The term *light* was originally restricted to radiation visible to the human eye, with wavelengths between 400 and 700 nanometers (nm). However, it has become customary to refer to radiation in the spectral regions adjacent to visible light (in the near infrared from 700 to about 2,000 nm) as *light* to emphasize the physical and technical characteristics they have in common with visible light.

link   A fiber-optic cable with connectors attached to a transmitter (source) and receiver (detector). In FSO systems a set of two terminals.

local area network (LAN)   A geographically limited communications network intended for the local transport of data, video, and voice.

loose tube   Type of cable design, primarily for outdoor use, where one or more fibers are enclosed in hard plastic tubes. Fibers are usually buffered to 250 microns, often filled with a water-blocking gel.

loss   Attenuation of optical signal, normally measured in decibels.

macrobending   Macroscopic axial deviations of a fiber from a straight line, in contrast to microbending.

material dispersion   *See* dispersion.

mechanical splicing   Joining two fibers together by mechanical means to enable a continuous signal. Elastomeric splicing is one example of mechanical splicing.

megahertz (MHz)   A unit of frequency that is equal to one million hertz.

mesh   Network architecture that provides redundancy.

microbending   Curvatures of the fiber that involve axial displacements of a few micrometers and spatial wavelengths of a few millimeters. Microbends cause loss of light and consequently increase the attenuation of the fiber.

micron (um)   Another term for micrometer. One millionth of a meter or 10^{-6} meters.

Mie scattering   Light scattering mechanism caused by particles the size of which is close to the transmission wavelength.

misalignment loss   The loss of power resulting from angular misalignment, lateral displacement, and end separation.

modal dispersion   Pulse spreading due to multiple light rays traveling different distances and speeds through an optical fiber.

mode   A term used to describe a light path through a fiber, as in multimode or single-mode.

mode field diameter (MFD)   The diameter of optical energy in a single-mode fiber. Because the MFD is greater than the core diameter, MFD replaces core diameter as a practical parameter.

mode filter   A device used to remove high-order modes from a fiber and thereby simulate EMD.

mode mixing   The numerous modes of a multimode fiber that differ in their propagation velocities. As long as they propagate independently of each other, the fiber bandwidth varies inversely with the fiber length due to multimode distortion. As a result of inconsistencies of the fiber geometry and of the index profile, a gradual energy exchange occurs between modes with differing velocities. Due to this mode mixing, the bandwidth of long multimode fibers is greater than the value obtained by linear extrapolation from measurements on short fibers.

mode scrambler   A device composed of one or more optical fibers in which strong mode coupling occurs. Frequently used to provide a mode distribution that is independent of source characteristics.

modulation   Coding of information onto the carrier frequency. This includes amplitude, frequency, or phase modulation techniques.

monochromatic   Consisting of a single wavelength. In practice, radiation is never perfectly monochromatic but, at best, displays a narrow band of wavelengths.

MQJ   Measurement Quality Jumper. A high-quality reference cable designed to provide accurate and consistent test results. The U.S. Navy requires that MQJs are used to test all Navy shipboard fiber installations.

multimode fiber   A fiber type that supports multiple light paths through its core. An optical waveguide in which light travels in multiple modes. Typical core/cladding sizes (measured in microns) are 50/125, 62.5/125, and 100/140.

multiplex   The combination of several signals onto a single communications channel.

multiplexing   The process by which two or more signals are transmitted over a single communications channel. Examples include time-division multiplexing and wavelength-division multiplexing.

NA-mismatch loss   The loss of power at a joint that occurs when the transmitting half has an NA greater than the NA of the receiving half. The loss occurs when coupling light from a source to fiber, from fiber to fiber, or from fiber to detector.

nanometer   A unit of measurement equal to one billionth of a meter, or 10^{-9} meters.

NEC   National Electrical Code. Defines building flammatory requirements for indoor cables.

numerical aperture   (1) A numerical value that expresses the light-gathering ability of a lens or fiber. (2) The imaginary cone that defines the acceptance area for the fiber core or a lens to accept rays of light.

optical fiber   *See also* fiber.

optical time domain reflectometer (OTDR) A method for characterizing a fiber wherein an optical pulse is transmitted through the fiber and the resulting backscatter and reflections to the input are measured as a function of time. Useful in estimating attenuation coefficient as a function of distance and identifying defects and other localized losses.

optical waveguide   Dielectric waveguide with a core consisting of optically transparent material of low attenuation (usually silica glass) and with cladding consisting of optically transparent material of lower refractive index than that of the core. It is used for the transmission of signals with lightwaves and is frequently referred to as fiber. In addition, there are planar dielectric waveguide structures in some optical components, such as laser diodes, which are also referred to as optical waveguides.

opto-electronic   Pertaining to a device that responds to optical power, emits or modifies optical radiation, or utilizes optical radiation for its internal operation. Any device that functions as an electrical-to-optical or optical-to-electrical transducer.

OTDR   Optical time domain reflectometer. A test instrument, working on the principle of continuous energy backscatter, which provides a complete characterization of fiber loss along its length.

patch panel   A centralized location for cross-connecting, monitoring, and testing telecommunications cabling.

PE   Abbreviation used to denote polyethylene. A type of plastic material used to make cable jacketing.

peak wavelength   The wavelength at which the optical power of a source is at a maximum.

photocurrent   The current that flows through a photosensitive device, such as a photodiode, as the result of exposure to radiant power.

photodetector   An opto-electronic transducer, such as a PIN photodiode or avalanche photodiode.

photodiode   A diode designed to produce photocurrent by absorbing light. Photodiodes are used for the detection of optical power and for the conversion of optical power into electrical power.

photon   A quantum of electromagnetic energy.

physical contact (PC)   Connectors that are aligned and mated so that no air gaps exist between them. Positive contact between fibers exists.

pigtail   Fiber-optic cable that has connectors installed on one end. *See also* cable assembly.

PIN photodiode   A diode with a large intrinsic region sandwiched between p-doped and n-doped semiconducting regions. Photons in this region create electron hole pairs that are separated by an electric field, thus generating an electric current in the load circuit.

plastic fiber   An optical fiber having a plastic core and plastic cladding.

plastic-clad silica fiber   An optical fiber having a glass core and plastic cladding.

plenum   (1) An air-handling space such as that found above drop-ceiling tiles or in raised floors. (2) A fire-code rating for indoor cable.

plenum cable   A cable whose flammability and smoke characteristics allow it to be routed in a plenum area without being enclosed in a conduit.

point-to-point   A connection established between two specific locations, as between two buildings.

polarization stability   The variation in insertion loss as the polarization state of the input light is varied.

preform   A glass structure from which an optical fiber waveguide can be drawn.

prefusing   Fusing with a low current to clean the fiber end. Precedes fusion splicing.

primary coating   The plastic coating applied directly to the cladding surface of the fiber during manufacturing to preserve the integrity of the surface.

pulse spreading   The dispersion of a signal with time as it propagates through the transmission media.

PUR   Polyurethane. Material used in the manufacturing of a type of jacketing material.

PVC   Polyvinyl chloride. Material used in the manufacturing of a type of jacketing material.

receiver   An electronic package that converts optical signals to electrical signals.

receiver sensitivity   The optical power required by a receiver for low-error signal transmission. In the case of digital signal transmission, the mean optical power is usually quoted in watts or dBm (decibels referred to 1 milliwatt).

reflectance   Light that is reflected back along the path of transmission, from either the coupling region, the connector, or a terminated fiber.

reflection   The abrupt change in direction of a light beam at an interface between two dissimilar media so that the light beam returns into the media from which it originated.

refraction   The bending of a beam of light at an interface between two dissimilar media or a medium whose refractive index is a continuous function of position (graded index medium).

refractive index   *See* index of refraction.

regenerative repeater   A repeater designed for digital transmission that both amplifies and reshapes the signal.

repeater   A device that consists of a transmitter and a receiver or transceiver, used to regenerate a signal to increase the system length.

return loss   *See* reflectance.

ring network   A network topology in which terminals are connected in a point-to-point serial fashion in an unbroken circular configuration.

rise time   The time it takes the signal output to rise from low levels to peak value. Usually measured from 10% to 90% of maximum output.

riser   (1) Pathways for indoor cables that pass between floors. It is normally a vertical shaft or space. (2) A fire-code rating for indoor cable.

scattering   A property of media that causes light to deflect from the media and contribute to losses.

scintillation   Also called heat shimmer. Turbulence-related phenomena that causes BER degradation in FSO systems.

sensitivity   For a fiber-optic receiver, the minimum optical power required to achieve a specified level of performance, such as a BER.

Signal-to-Noise Ratio (SNR)   The ratio of signal power to noise power. Measured in dB.

simplex cable   A term sometimes used for a single-fiber cable.

simplex transmission   Transmission in one direction only.

single-mode fiber   An optical waveguide (or fiber) in which the signal travels through its core. The fiber has a smaller core diameter.

SMA   A connector type with screw threads.

source   The means used to convert an electrical information-carrying signal to a corresponding optical signal for transmission by fiber. The source is usually a light-emitting diode (LED) or laser.

spectral width   A measure of the extent of a spectrum. For a source, the width of wavelengths contained in the output at one half of the wavelength of peak power. Typical spectral widths are 20–60 nm for an LED and 2–5 nm for a laser diode. The width of wavelengths in a light pulse, based on 50% intensity.

splice (1) A permanent joint between two optical waveguides. (2) A means for joining two fiber ends.

splice closure A container used to organize and protect splice trays.

splice tray A container used to organize and protect spliced fibers.

splicing The permanent joining of fiber ends to identical or similar fibers, without the use of a connector. *See also* fusion splicing and mechanical splicing.

splitting loss *See* coupling ratio.

ST Straight tip. A connector type with a bayonet housing that is spring-loaded.

star coupler An active or passive device where energy presented at an input port is distributed through several output ports.

star network A network in which all terminals are connected through a single point, such as a star coupler.

step-index Fiber-optical fiber that has an abrupt ("step") change in its refractive index, due to a core and cladding that have different indices or refraction. Typically used for single mode.

strength member That part of a fiber-optic cable composed of Kevlar aramid yarn, steel strands, or fiberglass filaments that increase the tensile strength of the cable.

tap loss In a fiber-optic coupler, the ratio of power at the tap port to the power at the input port.

tap port In a coupler in which the splitting ratio between output pods is not equal, the output port containing the lesser power.

tee coupler A three-pod optical coupler.

thermal stability A measure of insertion loss variation as the device undergoes various environmental changes.

tight buffer A type of cable construction whereby each glass fiber is tightly buffered by a protective thermoplastic coating to a diameter of 900 microns. A high tensile–strength rating provides durability, ease of handling, and ease of connection.

time-division multiplexing (TDM) A transmission technique whereby several low-speed channels are multiplexed into a high-speed channel for transmission.

topology The physical layout of a network.

total internal reflection The total reflection of light back into a material when it strikes the interface of a material having a lower index at an angle below the critical angle.

tracking Ability of an FSO system to follow movements of the installation platform.

transceiver An electronic device that has both transmit and receive capabilities.

transducer A device for converting energy from one form to another, such as optical energy to electrical energy.

transmission loss The total loss encountered in transmission through a system.

transmitter   An electronic package that converts an electrical signal to an optical signal.

tree coupler   A passive fiber-optical component in which power from one input is distributed to more than two output fibers.

turbulence   An atmospheric phenomena caused by temperature differences.

UL   Abbreviation for Underwriters Laboratories, Inc., the primary independent U.S. safety certification enterprise.

uniformity   The maximum insertion loss difference between ports of a coupler.

waveguide   Structure that guides electromagnetic waves along its length. An optical fiber is an optical waveguide.

wavelength   The distance between two crests of an electromagnetic waveform.

wavelength dependence   The variation in an optical parameter caused by a change in the operating wavelength.

wavelength-division multiplexing (WDM)   Simultaneous transmission of several signals in an optical waveguide at differing wavelengths.

WDM   Wavelength division multiplexer. A passive fiber-optical device used to separate signals of different wavelengths carried on one fiber.

INDEX

SYMBOLS

Hey, you've got enough worries.

Don't let IT training be one of them.

Get on the fast track to IT training at InformIT,
your total Information Technology training network.

 | **www.informit.com** | **SAMS**

■ Hundreds of timely articles on dozens of topics ■ Discounts on IT books from all our publishing partners, including Sams Publishing ■ Free, unabridged books from the InformIT Free Library ■ "Expert Q&A"—our live, online chat with IT experts ■ Faster, easier certification and training from our Web- or classroom-based training programs ■ Current IT news ■ Software downloads ■ Career-enhancing resources